MICROPROCESSORS IN OPERATIONAL HYDROLOGY

WATER SCIENCE AND TECHNOLOGY LIBRARY

MICROPROCESSORS IN OPERATIONAL HYDROLOGY

Proceedings of the Technical Conference on the
Use of Microprocessors and Microcomputers in Operational Hydrology,
Geneva, 4–5 September 1984,
organized by the World Meteorological Organization

Edited by

WORLD METEOROLOGICAL ORGANIZATION

Geneva, Switzerland

D. REIDEL PUBLISHING COMPANY

A MEMBER OF THE KLUWER ACADEMIC PUBLISHERS GROUP

DORDRECHT / BOSTON / LANCASTER / TOKYO

Library of Congress Cataloging in Publication Data

CIP

Technical Conference on the use of Microprocessors and
Microcomputers in Operational Hydrology (1984 :
Geneva, Switzerland)
Microprocessors in operational hydrology.

Includes index.
1. Hydrology–Data processing–Congresses. 2. Hydrological forecasting–Data processing–Congresses. 3. Hydrological stations–Data processing–Congresses. 4. Microcomputers–Congresses. 5. Microprocessors–Congresses. I. World Meteorological Organization.
GB656.2.E43T43 1984 551.4′0285 85-28143
ISBN 90-277-2156-4

Published by D. Reidel Publishing Company,
P.O. Box 17, 3300 AA Dordrecht, Holland.

Sold and distributed in the U.S.A. and Canada
by Kluwer Academic Publishers,
190 Old Derby Street, Hingham, MA 02043, U.S.A.

In all other countries, sold and distributed
by Kluwer Academic Publishers Group,
P.O. Box 322, 3300 AH Dordrecht, Holland.

Printed in The Netherlands

TABLE OF CONTENTS

INTRODUCTION

This volume contains the papers presented at the Technical Conference on the Use of Microprocessors and Microcomputers in Operational Hydrology, held in Geneva, 4-5 September 1984. The conference was organized by the World Meteorological Organization during the seventh session of the WMO Commission for Hydrology.

During the last few years, the availability of cheap reliable micro-electronic devices, such as microprocessors and solid state memories, has lead to their incorporation in measuring instruments for use in many fields, including operational hydrology. The characteristics of these devices, such as low cost, low power consumption, high reliability, and toleration of a wide range of environmental conditions make them particularly suited to applications in operational hydrology. Operational hydrological instruments have to be capable of working unattended for long periods, usually in remote sites, far from mains electricity supplies. Until recently only the simplest mechanical recorders, driven by clockwork, were considered suitable for these applications, but a number of water resources agencies have now undertaken field trials leading to operational applications of microprocessor or microcomputer based instrumentation.

In addition to the direct use of microprocessors in measuring instrumentation and data transmission systems, a number of users are using low cost microcomputers to make computing power more widely available, for purposes such as primary data processing in field offices or for dedicated applications such as on line real time forecasting systems. The introduction of micromputers will cause changes in operational procedures and require training for staff at all levels.

These new developments will have a great impact on all users of hydrological instruments and for this reason the technical conference was held concurrently with the session of the WMO Commission for Hydrology. The conference reviewed the use being made of these new devices so as to enable the experiences of pioneer users to be made more widely available.

The technical conference considered five main topics:

A. The use of micro-electronics in field instruments and sensors (processing sensor inputs, solid state recorders);

B. Microprocessor controlled data transmission systems (DCPs for satellite application, VHF radio, meteorburst, land lines);

C. Primary data procesing using microcomputers (Preliminary data screening in field offices, data storage and limitations, relation with mainframe based systems);

D. The use of forecasting models on microcomputers (Operational models, calibration requirements);

E. Integrated microcomputer based systems (Operational models, calibration requirements).

Authors were asked to concentrate on describing operational systems and to deal with the following aspects:

- Experience with hardware, including accuracy, reliability, costs and benefits;
- Implications for operational procedures;
- Training requirements;
- Software aspects, possibility of transfer of programs from mainframe systems.

The 22 papers from 14 countries and one international organization, give a good review of the state-of-the-art of the development and application of microprocessors and microcomputers in operational hydrology. It will be noted that reports of successful use came not only from the larger industrialized countries but also from smaller countries and from countries usually regarded as developing, indicating that these techniques may be used universally.

The thanks of the World Meteorological Organization are due to the authors and to their services for the time and effort they have taken in the preparation of these papers.

DISCLAIMERS

THE APPLICATION OF MICROCOMPUTER TECHNOLOGY TO OPERATIONAL HYDROLOGY IN CANADA

K. Wiebe*, B.E. Goodison** and W.I. Pugsley***

INTRODUCTION

The on-site capability of microprocessor controlled data acquisition systems to acquire, process, store and, as required, transmit data via available communication systems has increased the managers flexibility in the operation of the data acquisition programs. Probably of even greater importance is that the timeliness with which these data can be delivered to the user community is providing the impetus for a reevaluation of our ability to manage our water resources. Scientists too are increasingly making use of microprocessor controlled instrumentation to obtain measurements of hydrological processes which heretofore were only possible under laboratory conditions.

Canadian operational hydrology has already reaped benefits through the use of microprocessor controlled instrumentation. This gender of instrument has been used to measure discharge under extremely difficult and rapidly varying flow conditions. At unattended remote sites, data collection platforms (DCPs) in conjunction with satellite telemetry provide data that in frequency, format and timeliness are compatible with current water management data needs. Both of these types of microprocessor controlled instruments are discussed in detail with special emphasis on application, impact on the operation and action taken to assure reliability.

DCPs capable of collecting data from up to 16 sensors with a full range of signal outputs such as analog voltages, contact closure pulses and parallel as well as serial digital output provided the opportunity to conduct experiments on the practicality of collocating hydrometric and meteorological sensors. The results of these experiments are presented and the implications of operating a data acquisition network for more than one purpose are reviewed.

*Head, Hydrometric Methods Section, Water Resources Branch, Inland Waters Directorate, Environment Canada
**Research Scientist, Hydrometeorology Division, Canadian Climate Centre, Environment Canada
***Chief, Hydrometeorology Division, Canadian Climate Centre, Environment Canada.

World Meteorological Organization (ed.), Microprocessors in Operational Hydrology, 1-23.

The increasing use of DCPs and other automatic data acquisition systems has spurred a rapid development of specialized sensors based on microelectronic technology. Both the Canadian Water Resources Branch and the Atmospheric Environment Services are actively promoting this development work. Results so far include two new anemometers specifically designed for use at automatic stations and highly accurate incremental and absolute shaft encoders which convert shaft rotation on analog recording equipment to digital values compatible with automatic station input requirements. The availability of this specialized instrumentation in turn has led to unique applications. Some of these, including a digital snow depth sensor based on an ultrasonic ranging unit and microprocessor system, are described.

The capability of acquiring relatively large quantities of data on a variety of environmental parameters affecting hydrological processes has required that these data be processed, analysed and presented in formats appropriate to the user agency. The additional requirement that these outputs be presented in real or near real time has dictated the use of high speed processors to achieve this goal. For specialized requirements or specific goal oriented agencies the use of microprocessors for this purpose is proving to be a cost-effective tool.

In some cases documents cited in this text are HOMS components or proposed HOMS components. Where applicable, these are identified in the reference section. Also attached is a list of Canadian HOMS components which are relevant to the topics discussed but have not been specifically cited in the text of this paper.

Flow Measurement Instrumentation

The Canadian Water Resources Branch has developed, over the past several years, a fully automated microprocessor controlled moving boat flow measuring system (Environment Canada, 1984a). This method, originally proposed by the United States Geological Survey for the measurement of flow on medium to large rivers, consists of traversing a river with a suitable craft along a pre-selected path normal to the current while continuously acquiring data of; the resultant of the river and boat velocities at a depth of 1 metre, the angle of the current meter with respect to the cross-section line and, the river depth. These data are compiled and processed for pre-selected segments to produce average river velocity, average segment depth and segment width which are then combined to provide segment discharge. Segment discharges are then added for total river discharge.

Figure 1 is a block diagram presentation of the major components of the moving boat flow measuring system. Sensors include; a component current meter attached to an aluminum vane, which in turn is fixed to a stainless steel rod that rotates on two bearings within an aluminum tube mounted vertically at the front of, or amidships of the boat, an electronic compass attached to the top of the stainless steel rod which during initialization is set to read zero degrees when the current meter vane is parallel to the section line and, an internal digital depth sounder which provides the depth data required for the discharge computations.

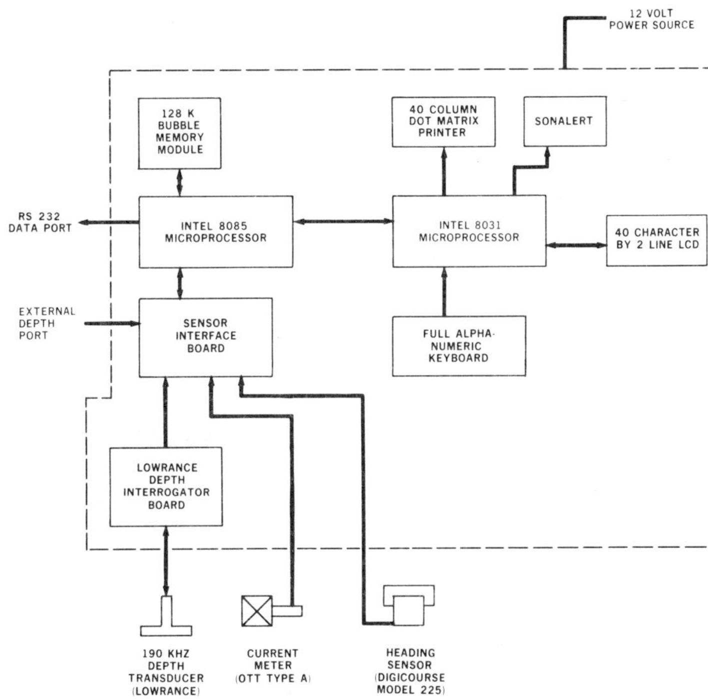

FIGURE 1 HARDWARE BLOCK DIAGRAM
DATEM MOVING BOAT DISCHARGE MEASUREMENT SYSTEM

The river velocity for each segment is computed from the sine relationship between the average velocity from the current meter and the average angle of the current meter relative to the section line as shown in Figure 2.

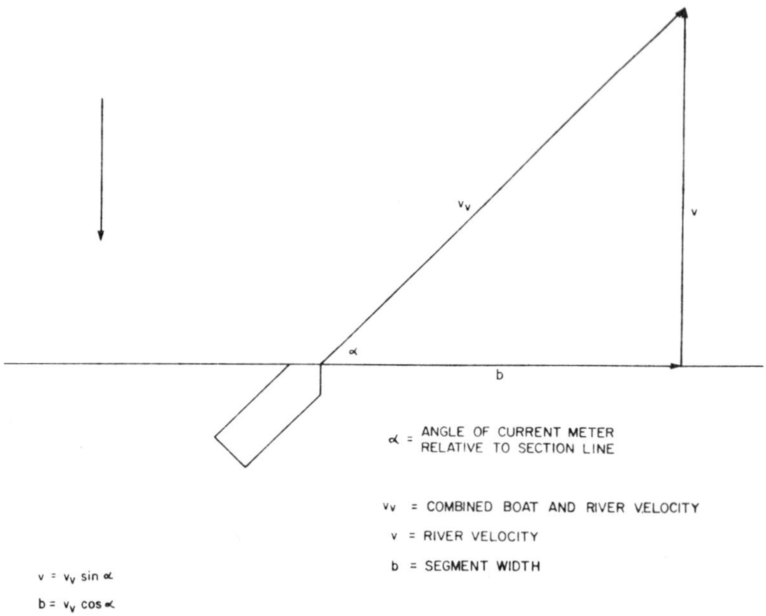

FIGURE 2 - RIVER VELOCITY AND SEGMENT WIDTH COMPUTATION

$$V = Vv \; SIN \alpha$$

where Vv is the average combined boat and river velocity

α is the average angle of current meter relative to section line

and V is the average river velocity

The segment width is provided by the cosine relationship

$$B = Vv \cos \alpha$$

where B is the segment width.

Site selection is very important to the success of a moving boat measurement. A poor site will effect the quality of data collected, as the technique is limited by the physical characteristics of a river. The following criteria should be used in selecting a measurement section;

i) River depths should be over 2 m in the traversing width. This minimum depth should extend at least 20 m upstream and downstream of the section line to avoid damage to subsurface equipment.

ii) Minimum river velocities should be over 0.5 m/s. The boat is difficult to maintain on course at river velocities below 0.5 m/s and errors due to excessive wandering off-line are induced.

iii) River sections with gently sloping banks should be chosen, as this will greatly assist section line targeting.

iv) A uniform cross section is preferred because a river with deep and shallow sections will be difficult to traverse as a result of variable current speeds.

v) River widths should be at least 100 m. Smaller widths will result in insufficient sampling points and hence in lower accuracy.

In addition to the above considerations, a measurement requires that the current meter equation, the ratio of the mean cross-section velocity to the mean velocity at the 1 metre depth point and, the exact river width be entered into the processor memory. The ratio of the mean cross-section velocity to mean velocity at the 1 metre depth point is obtained by conducting conventional velocity measurements at three to five verticals within the measuring cross-section. If the site exhibits steady uniform flow this type of calibration need only be carried out for certain stage segments, e.g. low, medium and high stage. The ratio of mean cross-section velocity to mean velocity at the 1 metre depth is typically in the 0.88 to 0.92 range if the velocity distribution in the verticals conforms to the Von Karman velocity distribution. If the river at the measuring site exhibits unsteady flow or flow under back water conditions, velocity measurements at selected verticals must be made at regular intervals for calibration purposes. Angular flow, providing the angle is not extreme, does not effect the accuracy of the measurement since a full discharge measurement consists of traversing the cross-section twice, once in each direction. River width is required for an overall accuracy check on the measurement and to make final correction. The processor checks the given width against the sum of the segment width and corrects for deviations.

In Canada the moving boat flow measuring system has been used on the Lower Fraser River and its major tributary in that reach, the Pitt River to calibrate a one-dimensional hydrodynamic model (Sydor et al , 1984). Results depicted in Figure 3 are typical of measurements under tidally induced unsteady flow conditions. River width on the Pitt River at the measurement site were approximately 300 metres and during the outflow portion of the tidal cycle, a seven hour period, a total of 26 complete flow measurements were made. Measurements were also made during the flow reversal portion of the cycle to provide data for complete cycle calibration. Note that a gap exists in measurements between the outflow and inflow portions of the cycle. During this period the flow velocities fell below 0.5 m/s and boat handling became too erratic for accurate measurements. It should be noted that conventional velocity measurements were made virtually continuously at three verticals within the cross-section to provide the required velocity ratio coefficient.

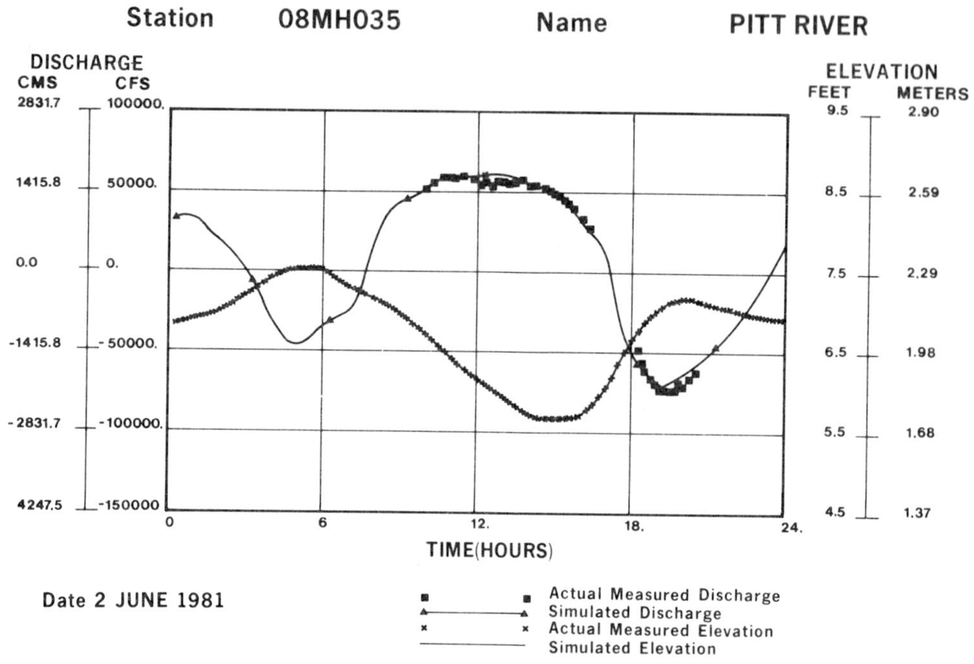

FIGURE 3 VERIFICATION OF ONE DIMENSIONAL MODEL BY MOVING BOAT TECHNIQUE

The system has also been used with considerable enthusiasm in the Canadian North (Environment Canada, 1984b). Here river flow conditions during spring break-up, when the major portion of the annual discharge of a river is delivered, can realistically be described as difficult. Flow at a measuring cross-section alternates between moving ice pans and short stretches of open water. To conduct a conventional measurement under these conditions is virtually impossible. With the moving boat system, however, measurements can be accomplished during open water periods without undue risk to either operators or instrumentation. During a two week period in the spring break-up of 1983, Water Resources Branch staff successfully established a rating curve on the Back River in the Northwest Territories. Had normal operating methods been used, the establishment of a rating would have taken several years. During this spring break-up (1984) another series of concentrated measurement programs using moving boat systems are being conducted on rivers in the Northwest Territories.

The Water Resources Branch has also carried out limited but successful experiments with acoustic velocity measuring systems. A system installed on the Niagara River during the early 1970's operated satisfactorily until ice movement and local development activity made the maintenance of transducer cables a full time occupation for field staff. Consequently, the site was judged uneconomical and was abandoned. On the Severn River in southern Ontario another acoustic velocity measuring system operated well for several years until struck by lighting. A second system subsequently installed has functioned well. However, problems have been experienced with the data recording units to which the system was interfaced. Current intentions are to interface with an automatic data acquisition system to permit the transfer of data via land-line communication.

Automatic Data Acquisition Systems

The Canadian Water Resources branch operates, under a cost share agreement with the provinces (excepting Québec), a network of approximately 2600 hydrometric stations. Québec in turn, through a cost share agreement with Canada, operates another 600 stations for an overall Canadian network of approximately 3200 stations. A current requirement is that 700 of these stations be equipped with automatic microprocessor controlled data acquisitions systems.

In Canada there are two basic reasons for installing this type of system at hydrometric stations. The first and most obvious reason is the requirement of real-time data from the site. The second is that the station is located at what is classified as a remote site. To be in the remote classification, a site must be accessible only by boat, helicopter or float equipped fixed winged aircraft. Excluding Québec, there are 350 sites with this designation as depicted in Figure 4.

To satisfy its requirement the Water Resources Branch has developed operational specifications to which these systems must be designed (Environment Canada, 1984c) and has actively sponsored system development. Current automatic data acquisition systems fall into two categories. These are: systems utilizing land-line communication networks and systems communicating via satellite.

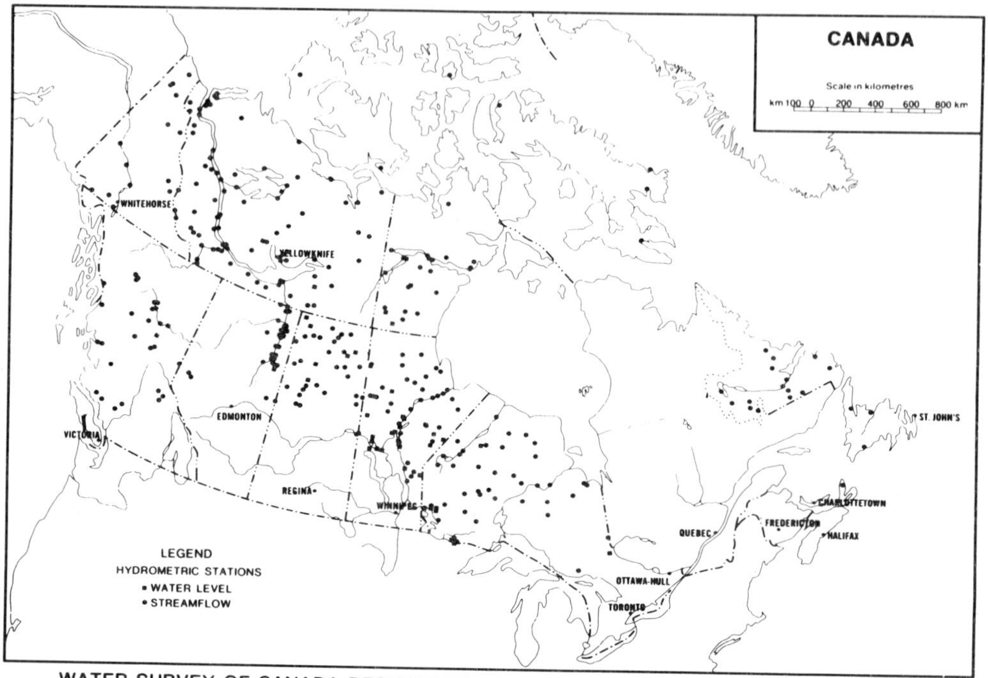

WATER SURVEY OF CANADA DESIGNATED REMOTE ACCESS HYDROMETRIC STATIONS

FIGURE 4

Not unexpectedly, systems utilizing land-line communication networks are found at readily accessible sites and fulfill the function of acquiring and transmitting real or near real-time data to the network operator. Generally, data requirements are specific and therefore these data acquisition systems are not designed for multi sensor operation. Typically, they acquire data from one or two parallel digital sensor inputs e.g. shaft encoders for water level and possibly an event input from a tipping bucket rain gauge. Since communication with the system is two-way, these systems are remotely programmable. The system currently in use can be programmed to acquire data at a given time interval or after the event monitored has changed by a predetermined amount or both.

Technically systems utilizing land-line communication networks are reliable. Operating temperatures for instruments with built-in modems (modulator-demodulator) are -40°C to +50°C. An exterior telephone company modem will not operate below 0°C and requires a heated instrument shelter. Data storage is provided electronically through RAM (Random Access Memory) with a capacity in excess of a month of data acquired at a rate of once every 15 minutes. Programming as mentioned earlier can be carried out either remotely using a standard terminal with an acoustic coupler (300 baud) or via a hand-held programmer connected to the instrument by means of an RS 232 data port. Program changes are processor prompted and response is numeric. The system operates on a 12 volt power supply and requirements are low. Quiescent power demands are in the milliamp range and daily consumption is in the order of 0.5 ampere-hour. At the majority of Water Resources Branch stations the power supply system comprises a photovoltaic panel, voltage regulator and 12 volt battery.

The utilization of systems using land-line communicatin networks, as mentioned earlier, is primarily for real-time data acquisition and data uses span the full spectrum of data needs in water resource management. Briefly they are; irrigation monitoring and control, navigation, water allocation, reservoir operation and flood forecasting. The use of this type of instrument to supply data for the latter two categories is not as high in Canada as the first three. There are two reasons for this. First, stations providing data for these management needs are generally located in remote regions of the upper watersheds. Second, instances during which the requirement for data becomes critical e.g. storm events, are those when land-line communication networks are the most vulnerable. Consequently, these sites are being equipped with data collection platforms (DCPs) which use satellite systems for data communication.

Since DCPs are used primarily at remote sites, travel to which is extremely costly, their design specification required that data acquisition be multi-purpose. Not only is the DCP required to provide the data necessary to keep the station manager informed on the status of the station but it also has to provide the capacity to permit the interfacing of sensors to monitor an ever expanding range of environmental parameters. Table 1 lists the parameters that the current DCP is capable of monitoring. Communication via satellite is one way. Therefore, the station managers cannot change a program without physically being at the station.

TABLE 1

PARAMETERS SENSED BY DOE HYDROMET DCPs

PARAMETER	SIGNAL OUTPUT	
Stage	Parallel Digital	Environmental Data
Temperature Water	Analog	
Temperature DCP	Analog	
Temperature Ambient	Analog	
Precipitation	Parallel Digital	
Rain Fall	Contact Closure	
Wind Speed	Parallel Digital	
Wind Direction	Parallel Digital	
Barometric Pressure	Analog	
Solar Insolation	Analog	
Ice Out Indicator	Contact Closure	
Ice Thickness	Analog	
Nitrogen Tank Pressure	Analog	
Battery Voltage	Analog	Station Performance Data
Recorder Clock Operation	Contact Closure	
Sediment Sampler Operation	Contact Closure	
Intrusion Counter	Contact Closure	
Emergency Signal	Contact Closure	

Technical specifications for DCPs are stringent. Since they must operate for long periods of time unattended, their design must be such that they can withstand shock and vibration due to transport, operating temperatures from -40°C to +50°C, temperature shock due to temperature variations of 15°C per minute, 100% humidity (non-condensing) and any electromagnetic radiation which they may encounter under normal transportation and operating conditions. The tests to which the DCPs are submitted are detailed in the following section. Transmission to the satellite is accomplished by a radio transmitter meeting the exacting specifications of the satellite communication system. Canadian DCPs predominantly communicate via the Geostationary Operational Environmental Satellites (GOES) which are operated by the National Environmental Satellite Data and Information Services (NESDIS). The GOES satellite data communication system consists of two operating satellites and an in-orbit spare. It makes up 2/5 of the World Weather Watch Global Observing System - satellite Sub-system. The other satellite used by the Water Resources Branch is the Tiros-N polar orbitin satellite operated by Argos in Toulouse, France. Data transmission via GOES satellites for Water Resources Branch DCPs is in the self-timed mode i.e. every three hour, the DCP has a 1 minute window within which it is permitted to transmit data. Consequently, data storage requirements are limited to about 350 characters and is archived through the use of RAM. Sensor interfacing is accomplished through a terminal strip interface. This method of interfacing, although slightly less reliable than plug connectors, was chosen to facilitate installation. DCPs are programmed by means of a hand-held programmer or a standard portable computer terminal via an RS 232 data port. The DCP processor prompts the programmer and entries are numeric or Y (yes) and N (no). The DCP requires a 12 volt DC power supply and obtains it from a photovoltaic panel battery power supply system. Typical daily usage is approximately 1 ampere-hour.

The primary reason that the Water Resource Branch has for equipping 350 remote hydrometric sites with DCPs is to provide greater management control over these stations. A study indicated that the Branch could save $1000 per annum per site in travel costs alone through the availability of real-time data from these stations. Increasingly, however, stations are being equipped with DCPs to provide data for water resource management purposes. In the order of numbers if not priority, DCPs provide data for reservoir operation, flood forecasting, navigation, water allocation and irrigation control. The other application of DCPs as will be mentioned later in this paper is in the operation of multi-purpose networks.

Quality Assurance for Data Acquisition Systems

To equip its 350 remote hydrometric sites with DCPs, the Water Resources Branch in 1983, embarked on a five year program for DCP acquisition and installation. Each annual DCP acquisition contract requires that DCPs are tested in the broad categories of contract qualification, acceptance and, reliability to ensure that each DCP meets the quality and reliability criteria of an operational unit.

The following is a description of test procedures;

A) Qualification Test Procedures.

These procedures require that one DCP of those manufactured under contract specification be subjected to mechanical shock and vibration at the level to be expected due to overland or air transport and to thermal shock, thermal cycle and humidity at the levels to be expected at any site in Canada. Further, the DCP is tested for radiation susceptibility at probable levels to be encountered in the field and also for radiation emission to ensure that they are not contravening federal regulations for the operation of such instruments if they are located near sensitive installations e.g. airports.

The design of the DCP shall be such that it will comply with the technical requirements of the DCP specification, (Environment Canada, 1984c) during and after exposure to test conditions stated above. The qualification test unit is tested under conditions which exceed requirements by a minimum of 10 percent.

B) Acceptance Test Procedures

These tests are required to ensure that strict quality control in the manufacturing procedures is being maintained. Each DCP is tested under the following simulated environmental stress conditions:

i) thermal cycle
ii) temperature shock
iii) vibration, and
iv) mechanical shock

C) Reliability Demonstration.

This test is carried out to determine the operating reliability of the DCP under conditions of maximum environmental stress. Since maximum stress on electronic instrumentation occurs at maximum operating temperature, the units are subjected to that temperature (55°C) when tested. The duration of the test is based on the number of units included in the test and is determined through the use of the curve in Figure 5 . Each unit must demonstrate successful operation both during and after completion of the test.

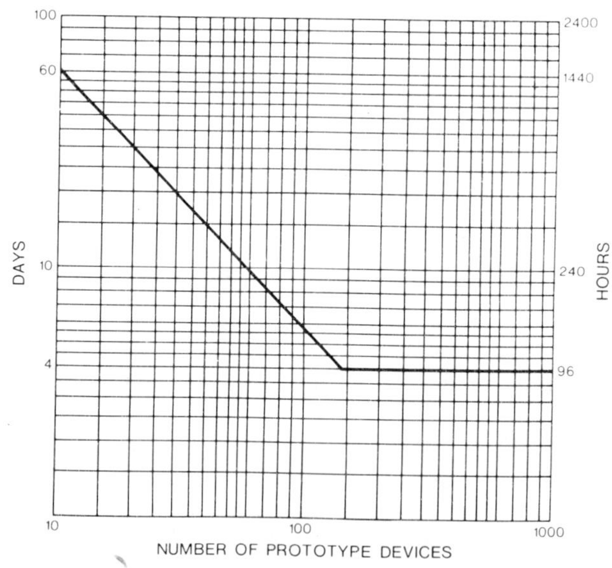

FIGURE 5 RELIABILITY DEMONSTRATION CURVE

Implications for Network Operation and Sensor Development

Microprocessor-controlled data acquisition systems could have a large impact on hydrometric and meteorological network operations in Canada. Responsibility for operation of such networks is divided among many federal and provincial agencies. Water Resources Branch (WRB) and Atmospheric Environment Service (AES) of Environment Canada are responsible for network operations on a national level. Although directed by the same government department, the hydrometric and meteorological observing programs are somewhat independent of each other. However, the expanded use of DCP's by the Water Resources Branch offers other agencies, such as AES, the opportunity to acquire field data from data-sparse, remote areas. The potential co-location of hydrometeorological sensors at WRB gauging stations could minimize field installation and operating costs. It was necessary, however, to develop and investigate operating strategies to make optimum use of this potential.

One of the first steps was the development of a policy on services for communication and archiving of data from automatic stations by the Atmospheric Environment Service, Environmental Conservation Service (Water Resources Branch) and the Canadian Forestry Service (Environment Canada 1983). The policy states the level of services that Environment Canada can offer to those agencies operating automatic stations for the acquisition of environmental data. It also specifies the required standards and procedures so that the data are compatible with those collected by the Department.

The policy contains the current specifications and standards for surface observation parameters of primary interest to AES and WRB from land-based automatic stations and data collection platforms. The associated accuracy specifications relate to climatological and hydrological applications of the data on a macro-scale. The standards for the hydrometric parameters of water level and water velocity and the atmospheric parameters of air temperature, precipitation, humidity, wind, solar radiation, sunshine, air pressure, and snowpack are contained in the policy. For each parameter, there are detailed specifications on the reported variables, data accuracy (target, minimum useable and normally achieved), sampling and reporting/processing requirements, siting and shielding sensors in current use and references.

The document also outlines the format and quality control standards for archiving data from automatic stations. Direct real-time capture of encoded data from the AES weather communication circuits is the most automated and most labour free in the maintenance stage; however, considerable software development is necessary. Development of the decoding and quality control routines for the meteorological parameters is being planned. Although micro-technology within AES has allowed for substantial change in sensors, on-site processors, data collection and analysis, the full benefits of micro-assisted technology will not be

realized until automated data management and archiving systems are available for network operations. Examples of how micro-computers are being tested in this role are discussed later in the paper.
As a further action to explore some of the advantages and problems associated with co-location of hydrometric and meteorological sensors, AES installed and operated meteorological sensors on a prototype WRB DCP. Details of the test are given in Goodison et al. (1983). Not only was the test designed to gain experience in operating a joint agency DCP station, but it provided a comparison of data from DCP sensors with those obtained from nearby standard synoptic or climate stations. Since the sensor, method of observation and siting at a DCP are generally different from those at standard climate and synoptic stations, there was concern about the compatibility of DCP data and standard station data. For example, if air temperature is sampled hourly on a DCP, the "true" minimum and maximum may not be measured as these may occur between hourly sample observations. Yet, the mean temperature may be similar. Generally, power limitations prohibit continuous observation of a parameter; the element is instead sampled at discrete intervals. The time of observation on a DCP may also differ from a standard observing station. Some DCP's sample on the hour or half-hour, some may make observations every three hours, while others sample just before their transmission time. In the latter case, one should not expect DCP observations to match exactly with nearby standard station observations. Users of data must make themselves aware of the method of observation and the potential effects on their analysis.

In the study noted here, measurements form the DCP located near the Ottawa River were compared with observations from a nearby synoptic station (Ottawa International Airport) and a standard climate station where observations are taken twice daily. Differences in meteorological data among the three stations could be a result of site, sensor or data abstraction. Daily data were compared for November 1982 and February 1983.

Precipitation storm totals were compared for the three stations for the test series. No bias was observed in rainfall or snowfall measurements at any of the stations; small differences in precipitation amounts were attributed to showery conditions or storm track differences. Hourly values of temperature and relative humidity were used to determine mean daily vapour pressure. The correlation of vapour pressure between the DCP and the AES stations was surprisingly good, considering one might expect higher humidity at the DCP site, which was located on the river.

Wind speed and direction proved to be much more difficult parameters to relate. Wind directions at the DCP site were quite different from those at the airport due to the sheltered location and river valley influence at the DCP site. Mean daily wind speeds at this site were consistently lower than the airport. This would be expected because of the sheltered location of the DCP platform.

Maximum daily temperatures at the DCP site and at the airport were very similar across the wide range of temperatures sampled during the November and February period (Figure 6). Minimum daily temperatures, on the other hand, tended to be as much as 5°C colder at Lemieux Island when minimum values fell below - 10°C (Figure 7). The difference is most likely due to a combination of cold air drainage into the river valley, which affects the DCP site, and/or local heating at the airport observation site, which was located in front of the terminal building.

Water level data were acquired over a period of time and found to compare favourably with those obtained by conventional means (Gagné, 1983).

This experiment demonstrated that AES meteorological sensors could be operated successfully on a WRB data collection platform . It also demonstrated that:

i) the DCP with its own complement of sensors and cables should be hooked-up and pre-tested prior to field installation;

ii) meteorological sensors should be sited to represent regional climate;

iii) precipitation (rain <u>and</u> snow) measurement is preferred to rainfall measurement only;

iv) technical training on Hydrometeorological DCP's and sensors is required ;

v) on-site testing and calibration of the DCP and sensors are necessary.

The development of reliable sensors and the design of special interfaces for connecting sensors to DCP's have progressed quickly. Microtechnology has been widely used in these technological improvements. AES has been involved in the design and development of several new meteorological sensors which employ micro-processors in their circuitry. The development of two new anemometers specifically for use at automatic stations is a case in point. Perhaps of more relevance to hydrology and hydrological operations, however, are developments related to precipitation measurement at automatic or remote stations. In Canada, of course, hydrologists are concerned not just with rainfall and rainfall intensity, but also with snowfall and snowcover.

The tipping bucket rain gauge, the standard for many years, is easily connected to DCP's or micro data loggers as an event input. There is no change in the cost of this sensor for use at automatic stations. As for long duration weighing precipitation gauges, AES has added an absolute position analogue to digital shaft encoder to measure the position and movement of the bucket. The shaft encoder requires no external memory. The output code is a true montonic function of the shaft position; the encoder outputs a series of unique binary words consisting of "1"'s and "o"'s that absolutely define every shaft position. Not every data logger

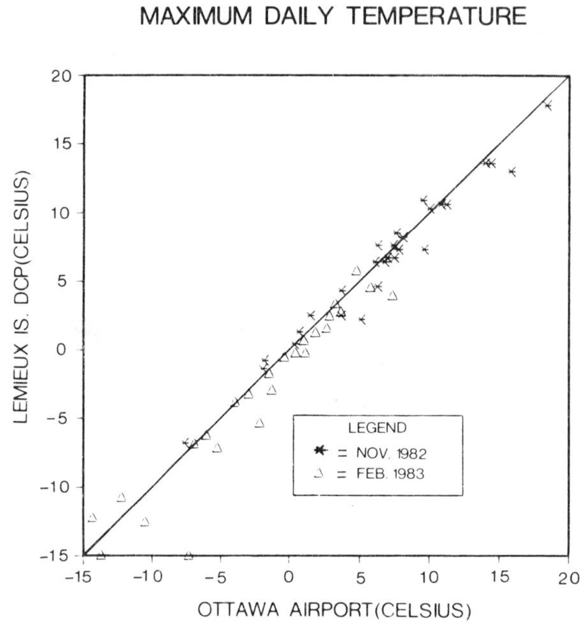

Figure 6. Comparison of daily maximum temperature at Lemieux DCP and Ottawa Airport.

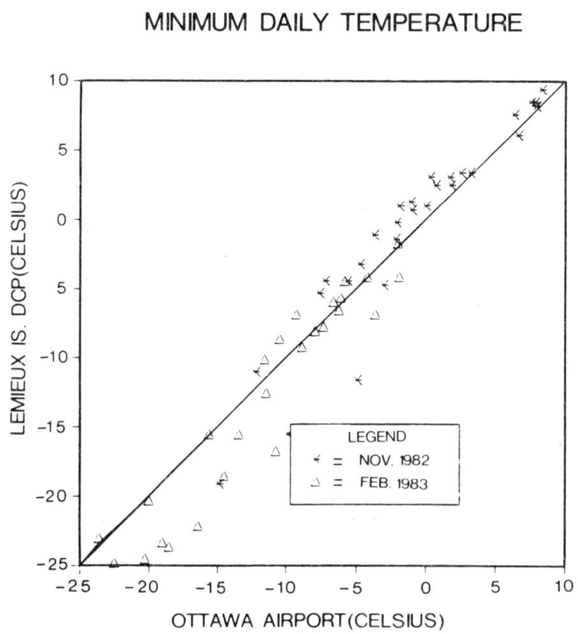

Figure 7. Comparison of daily minimum temperature at Lemieux DCP and Ottawa Airport.

can record the output; the Water Resources Branch's DCP accepts such an input and is programmed to output precipitation directly in millimetres. The shaft encoder currently used has proven accurate and reliable in Canadian conditions but, at a relatively high cost of about $2,500 CAN installed. This increases the cost of the long duration precipitation gauge used by AES to about $6,000 CAN. Without the encoder, however, the mechanical gauge could not be operated on a DCP and the possibility of acquiring precipitation data in real-time would be lost. The only alternative would be to use a tipping bucket rain gauge, which AES has chosen to do at many of its own auto stations, but this precludes the measurement of winter precipitation notably snow.

Tipping bucket gauges, however, are the basis for measuring rainfall intensity. For some studies such as urban stormwater runoff they may not provide data of sufficient accuracy, particularly if strip chart recorders are used and are manually abstracted. As an alternative, James and Haro (1983) reported the development of a precipitation instrumentation package for measuring rainfall intensity on a fine temporal and spatial resolution. The system is composed of a drop counter rain sensor, a microcomputer-based data acquisition system, an intelligent data decoder and a software package for the base microcomputer; it provides an automated, low-cost, high-resolution, reliable and intelligent system for rainfall data management, acquisition and presentation. The drop counter precipitation sensor is currently capable of measuring up to 170 mm of rain per hour. The data acquisition system uses a microcomputer with 64K bytes RAM. Forty of these are available to store up to 20 data sets while the remainder is used for program execution. When the memory space is full, the microcomputer dumps the block of information onto a cassette recorder. The integration time is programmable and the number of data sets recorded depends on the duration and not the amount of rain. For a one minute integration period, the system can be left unattended for 7 days of continuous rain or more than 4 years of no rain. A microcomputer based data decoder retrieves the data from the cassette tape and transfers it to a central computer for data processing.

A more detailed description of the system is given in James and Haro (1983). Comparison of data from this system with that of a tipping bucket rain gauge shows the former to be more sensitive, giving a better indication of the beginning and end of rainfall, yet with comparable accuracy as to amount. This is extremely useful for application to urban stormwater management. One problem, yet to be resolved, is protection of the data acquisition system from strong electromagnetic effects in hostile environments. This is a problem which is common to many data logger systems. Another problem recently encountered is that the upper limit for rain rate could be exceeded over short intervals. This is currently being investigated.

As noted previously, the capability of measuring both rain and snow is necessary in Canada. The automation of snow measurement at unmanned stations is in fact much more difficult than for rainfall only. One method being investigated by AES is the development of reliable, low-cost automatic snow depth sensor which could be used at unmanned meteorological and hydrometeorological stations. The sensor system being developed is based on the principle of ultrasonic wave reflection in air. Figure 8 provides an overview of the ultrasonic snow depth gauge system.

FIGURE 8 - ULTRASONIC SNOW-DEPTH GAUGE

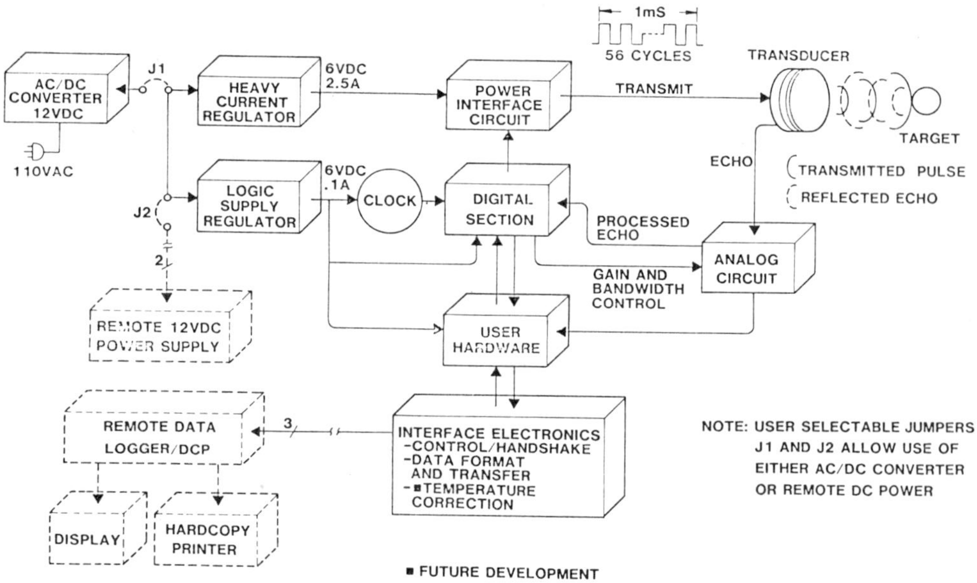

A commercially available ultrasonic ranging kit from Polaroid Corp. provides an economical means of accurately measuring target distance in terms of time to target. An on-board crystal oscillator provides a time base pulse train which is activated on a measurement request command from an external device such as a data logger or data collection platform (DCP). Once the pulse train is enabled it up-counts designated registers until a target echo is received. This immediately disables the time base pulse train. The numbers memorized in the counter can now be multiplied by the precise oscillator pulse period to give an accurate round trip time from transducer to target and return. Dividing by two gives the required time to target. The target distance is determined by multiplying the time by the velocity of sound in air at a known reference temperature. The calculated target distance is then subtracted from the actual measured distance between the sensor and a reference target (i.e. the ground at the site). Since the velocity of sound in air is highly temperature dependant, a correction must be applied to adjust the velocity for ambient air temperature. Measurement data are then formatted for transmission to an appropriate peripheral for display and/or archive. A microprocessor and temperature sensor built into the system would allow for real-time temperature correction and provide flexibility for formatting data and subsequent transmission to remote devices such as data loggers and DCPs.

The sensor developments noted above depend very much on current microprocessing capabilities. Problems related to the interfacing of sensors to DCPs and data loggers are being resolved. Of greater concern to users now is the development of quality control procedures, effective data management and formatting of data from such systems that would permit the exchange of real-time hydrologic date between users. The advantages of real time data collection and processing developed through the effective application of microprocessors or microcomputers will be lost without efficient data management.

The Expanding Role of Microcomputers in Data Collection Processing and Analysis

The rapid expansion of DCP networks, combined with the desire to acquire data from these stations in a timely and cost-effective manner, has led to the development of data collection systems tailored to an individual agency's needs. In Canada, no water resource agency has installed a direct-readout ground station to collect real-time environmental data from its automatic stations using GOES satellite telemetry. Instead, agencies access the NOAA/NESDIS (National Environmental Satellite Data and Information Service) computer in Camp Springs, MD, USA, using dedicated or dial-in telephone circuits. Environment Québec accesses data from 90 DCPs operated throughout the province using a direct telephone link-up to NESDIS to carry data back to its Québec City office. There, a minicomputer (internal memory of 256 K-bytes, disc storage of 22.5 megabytes) carries out a number of functions, including automatic collection of data, monitoring of the quality of messages received, temporary storage of data for up to seven days and daily transfer of data to a main-frame computer (C. Pesant, 1984). The system can also disseminate processed data on request.

Other agencies in Canada are in the process of developing or implementing data collection systems to retrieve messages transmitted through GOES satellites and to process and disseminate the data. Agencies with large DCP networks require automatic data retrieval; manual data retrieval by means of a dial-up computer terminal is sufficient for a network of less than 10 stations. The Saskatchewan Research Council in conjunction with the Canadian Climate Centre has developed a microcomputer based data collection system for retrieving and processing DCP data. A microcomputer (256 K RAM, 360 K floppy disc drive and 5 megabyte hard disc) with an auto-dialer has been programmed to access on a daily basis the NOAA/NESDIS computer, retrieve the DCP data on file, store the data in raw form and then automatically print out the processed data in tabular form. This system eliminates the need for a person to dial repeatedly, is programmed to dial during the night when long distance telephone rates are cheaper, and provides a hard copy data summary (including messages about missing data) which can be quickly reviewed the next morning for data errors, data losses, DCP malfunctions or lost transmissions and NOAA/NESDIS system malfunctions. More detailed quality control routines for meteorological data are planned for development and implementation on this system in late 1984. These will meet the format and quality control standards for archiving data form automatic stations specified in the national policy on services for communication and archiving of data from automatic stations (Environment Canada, 1983).

With the increasing number of DCP stations and users, it is becoming more diffcult to access one's DCP data from the NESDIS computer using the direct dial-in method. The auto-dial approach (which can repeatedly dial a number or wait for a period of time before re-dialing) is one method of minimizing the effect of this problem. A dedicated circuit is another alternative. Atmospheric Environment Service (AES) of Environment of Canada can access their DCP data hourly using the National Weather Service circuit; however, a poor communication link between Washington and Toronto has resulted in data loss as high as 50%. This certainly detracts from effective operational use of data on a near real-time basis.

As noted above, microcomputers are being used more for processing of data. AES is currently implementing a microcomputer system to process punch-paper tape records from 150 long duration digital recording precipitation gauges. The previous processing system, which has been in place since 1969, uses an automatic translator which transfers data from the punched-paper tape recorded on site to punch cards which then serves as input to a main-frame computer.. This system is described by Pollock et al., 1973. The new system will basically perform the same functions as the previous one; however, data will no longer have to be transferred to punch cards. The microcomputer (40 K bytes RAM, 256 K bytes RAM, 360 K bytes floppy disc drive, 10 megabyte hard disc) is interfaced to the translator and currently supplies data to the existing processing and quality-control programs on the main-frame computer from which interim tabulations and an archive of fifteen minute precipitation totals is created. By late 1984, updated processing and quality-control programs will be transferred to the microcomputer, on which all processing will be conducted. The system is designed to handle 100-150 days of precipitation data (96 readings per day) at a time. The main-frame computer will only handle the data for the digital archive. Reasons for implementing a microcomputer system include the phasing out of punch card support in the current data processing system, the desire to eliminate the time consuming procedure of punch cards, and the need to update the quality-control routines. The microcomputer based system offers the capacity and flexibility for performing the required tasks.

Microcomputers are now widely used for analysis of hydrometeorological data. One unique application has been the development and operational testing of a procedure for mapping basin snow cover based on a microcomputer interfaced with a low-cost video-image analyzer. Snow cover maps for the Saint John River Basin in eastern Canada have been prepared from visual NOAA satellite imagery using subjective photo-interpretation techniques (Schneider et al., 1976) and from visual and near infra-red data, using an objective image analysis system (Waterman et al., 1980). Recently, a procedure was developed to map snow cover for the Saint John basin and its 70 sub-basins from either NOAA or GOES imagery using a low-cost video-image analysis system, interfaced to a microcomputer.

Specific details on this procedure are given in Johnstone and Ishida (1984). The basic equipment includes the video-image analyzer (49,152 bit memory; light pencil) television monitor, video cassette recorder, television camera, and a microcomputer (including compatible disc drive, monitor, keyboard, 64 K bytes RAM). Data input was from an illuminated photograph of the satellite image by means of a television camera. The image was enlarged to overlay the Saint John River Basin base map used in the study; each bit in the image analyzer corresponded to about 4 km2. A bit map of geographical reference points was created for use in the analysis. Different base maps were needed for analysis of polar orbitting (NOAA) and geostationary (GOES) satellite imagery. A microcomputer was used to plot and digitize basin and sub-basin boundaries suitable for NOAA imagery while a zoom transfer scope was used to transform the basin boundaries into a perspective equivalent of GOES-E. Both base maps were loaded into the image analyzer as a bit map and then stored on a diskette for future use.

The method of analysis for snow cover determination is described by Johnstone and Ishida (1984). This report also includes a comparison of this method of snow cover mapping with the other methods noted above for the Saint John Basin during the 1982 and 1983 melt seasons. It was concluded that the image analyzer/microcomputer system was generally capable of providing an estimate of the total basin snow cover and a snow cover map comparable to the results obtained by the AES digital technique and subjective NESDIS method. Three main problem areas currently remain: availability and suitability of the data in a hard copy format for application of this technique at a centre far from the satellite reception point; deficiencies in the analysis system; and, inexperienced analysts. The technique has been demonstrated to be an operational success and the technology has been transferred to local consultants in the private insdustry for direct operational application to the Saint John Basin.

References

Environment Canada, 1983: Policy on Services for Communication and Archiving of Data from Automatic Stations. Atmospheric Environment Service, Downsview, Ontario. (Proposed HOMS Component).

Environment Canada, 1984a: Hydrometric Field Manual - Automated Moving Boat Measurement System. Inland Waters Directorate, Water Resources Branch, Ottawa (In Press). (Revision of HOMS Component E79.2.06 Discharge Measurement by Moving Boat Method.

Environment Canada 1984b: Northwest Territories Hydrometric Station Investigation Report for 1983. Inland Waters Directorate, Water Resources Branch, Yellowknife.

Environment Canada 1984c: DCP Specifications. Inland Waters Directorate, Water Resources Branch, Ottawa. (Proposed HOMS Component).

Gagné, A.D., 1983: Hydrologic Data Collection Platform Reliability. Internal report, Environment Canada, Water Survey of Canada, Ottawa. 56 pp.

Goodison, B.E., Wilson, B., Wu, K., Metcalfe, J., 1984: An Inexpensive Remote Snow-depth Gauge: An Assessment. Proc. Western Snow Conference, Sun Valley, Idaho, April 17-19, 1984 (in press)

James, W., and Haro, H., 1983: An Economic Automated System for Acquisition of Precipitation Data. Paper presented at Inland Waters Directorate Workshop on Streamflow Forecasting, Toronto, Ontario, November 1983. 11 pp.

Johnstone, K.J., and Ishida, S., 1984: An Analogue/Digital Procedure for the Mapping of Snow Cover. Canadian Climate Centre, Report No. 84-9 Atmospheric Environment Service, Downsview, Ontario 28 pp.

Pesant, C., 1984: Telemetry - an Indispensable Tool to Provide Timely Data and to Improve Monitoring of a Hydrometeorological Network. National Lecture Tour 1984, Canadian Associate Committee on Hydrology, CSCE, 73 pp.

Pollock, D.M., Rogalsky, J., and Carr, D.A., 1973: A Processing System for Fischer and Porter Precipitation Gauge Data. CLI-6-73, Atmospheric Environment Service, Downsview, Ontario. 16 pp.

Schneider, S., Wiesnet, D. and McMillan, M., 1976: River Basin Snow Cover Mapping at the National Environmental Satellite Service. NOAA Technical Memorandum NESS 83, US Dept. of Commerce, Washington, D.C.. 19 pp.

Sydor, M. and Fast, E.J., 1984: Résultats d'une simulation en temps réel à partir d'un modèle hydrodynamique unidimensionnel et d'un système de mesure par bateau mobile appliques sur le fleuve Fraser. Présenté au Colloque de l'A.C.F.A.S., Ville de Québec. Mai 1984, pp 21.

Waterman, S., Hogg, W., Hanssen, A., and Polavarapu, V., 1980: Computer Analysis of TIROS-N/NOAA-6 Satellite Data for Operational Snow Cover Mapping. Proc. Sixth Canadian Symposium on Remote Sensing, Halifax, Nova Scotia, May 21-23, 1980. p 435-442. (HOMS Component D00.0.07 Snow Cover from Multi-channel Satellite Data).

Relevant Canadian HOMS Components

D00.0.06
Practical Methods of Aiming Antennae at Geostationary Satellites.

F00.2.11
Data Transmission Via Geostationary Satellites.

F00.2.12
Data Transmission Via Polar Orbiting Satellites.

J74.2.02
One-Dimensional Hydro Dynamic Modelling.

MICROPROGRAM RECORDING OF HYDROLOGICAL DATA ON READ-ONLY MEMORY

J.M. FRITSCH* and P. MAILLACH**

1. INTRODUCTION - GENERAL CONSIDERATIONS

The search undertaken by us for new data collection instrumentation in the field of hydroclimatology was based on the following objectives:

(1) To provide data acquisition on a machine-readable medium by means of electronic processing, thus eliminating the manual or assisted processing of graphic recordings which has always been a difficult area in hydrological studies.

(2) To improve data collection reliability.

Graphic recorders are very sensitive to environmental characteristics and many devices and components, such as clockwork mechanisms, paper, styluses, pens, etc., are likely to be adversely affected. There is good reason to believe, however, that a well-designed electronic system will provide better performance.

(3) To store significant parameters only.

While access to a programmable microprocessor enables the sensors to be scanned almost constantly, the storage command is only triggered if the parameter variation in question differs significantly from the most recently stored data.

(4) To reduce costs. The cost of precision engineered and clockwork sensors and data acquisition control systems, which are only manufactured in limited numbers, can only increase in the current economic climate. On the other hand, microprocessor-based systems are already less expensive than these devices, and it may be assumed that the price difference between mechanical and microporcessor-based systems can only increase as time goes by.

This article describes two systems designed with to the above-mentioned objectives in view: the CHLOE hydrological processor (Centrale HydroLogique ORSTOM-ELSYDE) using silicon sensors and a sophisticated processing unit; and an OEDIPE (Organe d'Enregistrement Digital de l'Information PluviométriquE) raingauge, with mechanical sensors (tilting bucket) and a simple processing unit. The OEDIPE system is available at a modest price and is suitable for use by non-skilled personnel.

* Research worker at ORSTOM, P.O. Box 165, 97323 Cayenne-Cedex, French Guyana
** Software Supervisor, ELSYDE, 95 route de Corbeil,
91700 Ste. Geneviève-des-Bois

World Meteorological Organization (ed.), Microprocessors in Operational Hydrology, 25-37.

Both systems use the recording medium and the data rereading unit described below:

2. CE64 MEMORY CARTRIDGE

The CE64 cartridge is a removable unit connected to the data acquisition system by a 96-pin connector. Its external dimensions are 115 mm x 115 mm x 20 mm.

The cartridge has a capacity of 64K bytes of programmable memory which may be erased by ultraviolet radiation (EPROM). This capacity is obtained through the use of eight chips of 64K bits each with standard 28 pin contacts (INTEL 2716, for example).

The primary advantage of the EPROM is the absence of an electric current to maintain the information stored in the cartridge, which makes this medium particularly solid and reliable.

In this technology, when the required bits are written to they are electrically set to zero. A return to the initial state is no longer possible unless energy amounting to 15 Watts per second per square centimetre is applied to the chip in the form of radiation of 250 nanometre wavelength supplied by an eraser lamp.

Erasure time is between 5 and 40 minutes depending on the position of the chip in the eraser tray. Once this operation is completed, all bits in the memory are reset to one.

The number of write-erase cycles that can be undertaken by a package is extremely variable, according to the various manufacturers. Some claim 1,000 cycles, others only 50. We therefore performed a series of bench tests on EPROMS in the CE64 cartridge and over 200 write-erase cycles were performed without damage to the memories.

In fact, the problem most likely to occur after a while is that some bits or bytes in a chip may jam instead of switching back to the zero state after a write instruction. The procedure used with RAM memories (new attempt at writing if the re-read information does not conform to the source information) cannot be used in EPROM technology - EPROM is a one-time shot! It would also be unwise to try to flag the next byte by an appropriate error code (which would enable the laboratory to invalidate the defective byte during read time) because it is quite possible that an entire zone adjacent to the memory, or even the whole 8K byte package, is defective. This considerable restriction has sometimes been cited as a major drawback when using EPROMs for data storage.

In the CHLOE and OEDIPE systems, we employ a special processing and reading procedure by using a microcomputer which enables us to clearly identify the defective bytes which were abandoned at write time. Consequently, the data acquisition system will continue to function, even if one or several packages are out of order. The effective capacity of the cartridge will be reduced, but the validity of the stored data will remain unaffected.

3. READING DATA BY USING A MICROCOMPUTER

CE64 cartridge read-outs in the laboratory are performed using a cartridge memory reader. A microprocessor is used for this purpose, capable of performing a conversational procedure with any micro- or mini-computer equipped with an asynchronous communications port, such as the RS-232C.

When this unit identifies a command of the "C AAAA FFFF" type from the computer, the cartridge contents included between the hexadecimal addresses AAAA and FFFF are transmitted to the processing system by means of the communications port at the rate of 4800 baud, in accordance with the ASCII code generally used in microcomputer data processing. The cartridge contents are also directly interpretable on a display console. By using the CHLOE system, the information stored on the cartridge is coded in BCD (four bits per character) and 128K characters will therefore be transmitted during the entire read-out operation of the cartridge.

This results in the transmission of 10 bits per character: 1 start, 7 data, 1 even parity, 1 stop. In addition to this transversal parity bit at byte level, the cartridge memory reader generates a longitudinal parity byte - an exclusive OR from all transmitted bytes. This (BCC) parity byte is transmitted at the end of each data stream, thus ensuring for the user virtually total security with respect to the conformity between data received and data stored on the CE64 cartridge.

An additional feature provided by the cartridge memory reader enables the user to ensure that the cartridge is completely empty after erasure by the ultraviolet rays. A push-button triggers the sequential reading of the 64K bytes by the cartridge memory reader microprocessor and a light-emitting diode blinks if an erasure defect is encountered. This feature is designed to avoid taking full or partially erased cartridges out into the field, only to be rejected by the data acquisition system.

The cost of a CE64 cartridge is approximately US$ 200. The read and cartridge memory reader test unit unit costs arouund US$ 650 (prices exclusive of tax, 1983).

4. THE OEDIPE RAINGAUGE

The function of the OEDIPE system is to store on the CE64 cartridge the times and dates of electrical impulses triggered by the tipping-bucket raingauge.

The OEDIPE system is housed in a (standard IP 569) plastic, impervious package which weighs approximately 5 kilogrammes and measures 350 mm x 250 mm x 150 mm.

The unit is equipped with a four-socket junction which connects the power lines and the lines linked to the mercury contact, which in turn is linked to the recording raingauge buckets. A light-emitting diode indicating the activation of the cartridge (read or write) completes the facade. This "black box" structure has been retained in order to simplify implementation procedures as much as possible and also to reduce manufacturing costs.

The recorder is automatically activated when a cartridge is inserted. The system is switched on and the microprocessor performs the following initiation programme:

- Reading of the 64K bytes in order to check that the CE64 cartridge is empty (all bytes are at 'FF');

- Writing of the recorder's serial number (factory-set) onto the first two bytes in the cartridge.

This sequence lasts approximately 30 seconds, during which time the 'cartridge activated' light-emitting diode is lit. If the procedure is successfully completed, the LED goes out. If the procedure is not successful a continuous blinking alerts the operator to the fact that the initialization procedure has failed.

This malfunction may be due to two possibilities:

(a) The cartridge is not 'clean';

(b) The write operation of the recorder serial number was not possible and the cartridge is defective.

The electrical current to the system is cut when the cartridge is removed.

When the cartridge is inserted, the OEDIPE internal time reference is initialized to zero. Each tilting motion will be referenced in time relative to this original instant. The user is responsible for noting the start-up date and time so that a recording raingauge card referenced in absolute time may ultimately be created. This procedure takes place in the laboratory when the cartridge is processed.

The temporal coordinates of each tilting motion are immediately stored in binary on four bytes in the following format:

Day of the month (day number)	10 bits
Hour	6 bits
Minute	8 bits
Second	8 bits

Using this coding, each CE64 cartridge can record 16,383 tilting motions, which correspond to the following 'climatological changes':

- 8191.5 millimetres of rain; 1 tip for every 0.5 mm rainfall (buckets 20 cm^3; funnel 400 cm^2);

- 1638.3 millimetres of rain; 1 tip for every 0.1 mm rainfall (funnel 2,000 cm^2).

The supply voltage required may vary between 15 and 11.5 volts DC. Because of the CMOS technology involved, the OEDIPE system only consumes 3mA at 12 volts at rest. During one write cycle, i.e. at each tipping motion, the current required is 400 mA for a period lasting approximately one second. Thus, monthly consumption - at the rate of 2,000 tipping motions per month - may be as low as 2,5 Ah. Electrical independence may thus be achieved by using rechargable batteries, mini solar panels or even dry cells.

It is planned that OEDIPE systems will be inspected and checked while on service in the field. This operation will be made possible by using a pocket mini-terminal, equipped with an LCD display which communicates the most recent OEDIPE recordings to the user, that is:

- The number of tipping motions recorded;
- The amount of memory used;
- The number of defective bytes encountered;
- The time and date of the most recently recorded tipping motion;
- Elapsed time since start-up;
- Battery voltage.

These data may be sent by a transmitter in the ARGOS satellite data-collection system. In this way the OEDIPE functions can be monitored by remote control and cartridge changes in far-off sites with difficult access can be optimized.

The cost of an OEDIPE recorder is approximately US$ 550 and the inspection terminal costs around US$ 600 (prices exclusive of tax, 1983).

5. THE CHLOE MULTI-PARAMETER HYDROLOGICAL RECORDER

Contrary to the OEDIPE system, which is equipped with a mechanical sensor and relatively simple software, the CHLOE system comprises an underwater (SPI) unit, equipped with silicon sensors, a microprocessor, program storage, input-output interface, a veritable underwater data processing system, and an aerial check (SET) unit, the face of which is composed of a keyboard with 12 watertight push-buttons, two 2-digit wheels, one 4-digit dip-switch and an LCD display (Figure 1). It is therefore a more sophisticated system than the OEDIPE and is capable of measuring continuously varying parameters. Setting up the CHLOE system is more complex than for the OEDIPE system.

5.1 THE UNDERWATER PIEZO-RESISTIVE TRANSDUCER (SONDE PIEZO-RESISTIVE IMMERGEE): SPI

5.1.1 Functions and thermal characteristics

The primary function of the SPI system is to measure the true value of the water level of a river in the site's reference hydrological system (set of staff gauges).

This information is obtained by differentially measuring the hydrostatic pressure and the atmospheric pressure. The SPI also measures temperature. Moreover, while testing the prototype in French Guyana, the density of water was measured (using the so-called SPI2 version), as well as the conductivity (using a supplementary probe - the PONSELLE-CTS10.

A piezo-resistive sensor is used to measure pressure. The sensor is made up of a silicon strain gauge which is implanted into a metallic membrane and which deflects under the pressure of the fluid. The deflection causes changes in the internal resistance of the piece of silicon, which causes the voltage across it to vary when subjected to a constant electric current.

This voltage, converted into a frequency, is analyzed by the microprocessor and converted into pressure, then into water level on the staff gauge using the pressure-frequency curve which is specific to the sensor and is established by calibration. The sensitivity of the procedure is such that, by using a sensor capable of supporting 1.5 bar (c.15 metres of water), a 0.1 pressure difference of millibar (i.e. approximately 1mm water) is identifiable at the frequency metering.

5.1.2 Thermal compensation for electronic components and sensor

The SPI system is designed to operate in a fairly variable thermal environment (0 to 50°C). Unfortunately, however, temperature variation has the same effect on silicon as the deflection caused by pressure. It therefore generates irrelevant parasitic information. Similarly, the analogue electronics which formulate the signal for the microprocessor (amplification and frequency conversion) vary significantly with temperature changes. However, thermal effects on the SPI digital electronics are negligible.

Two distinct influences need to be corrected with respect to the sensors and the analogue electronics.

Temperature correction in sensors

The piezo-resistive sensors used comprise various silicon chips which are fixed to a membrane and wired together in a Wheatstone bridge. The temperature variation effects, which in theory are identical on all gauges, is cancelled. In fact, despite these precautions, the gauges frequently remain unmatched after assembly. When static, this causes an imbalance in the bridge and, when in motion, a non-linear temperature drift. Manufacturers suggest on the calibration card of each sensor that several resistance values be planted in either branch of the bridge, in series or in parallel with the gauges, in order to compensate for these imbalances as much as possible. If the imbalance is still not corrected, it is corrected using software, by programming the results of a laboratory calibration in which temperature and pressure were varied. For a given temperature, the output frequency of the sensor is taken every 20 mb between 0 and 1000 mb for temperatures varying at 10° intervals.

This group of curves is stored as a table, in the probe's software, which makes the necessary interpolations.

During the calibration process, only the sensor is submitted to temperature variations in the container because the electronic processing card remains at a constant temperature.

Temperature correction on the electronic processing equipment

Compensation for the drift of electronic components is performed both in hardware and software in accordance with the auto-calibration principle of the components: two ultra-stable reference sources simulate the sensor responses corresponding to the two extreme pressure limits to be measured. The discrepancy between the frequencies transmitted by the two sources compared to those obtained at calibration temperature enables the data processing program to supply the required corrections to the signal as a linear correction.

Figure 2 presents a section of a calibration curve for two sensors, C1 and C2. Along the H = f(K) curve, which gives the water level as a function of the output frequency for a given temperature (18.7° for C1 and 21.5° for C2), the discrepancy is represented in Hertz for two temperatures flanking the calibration temperature (K = f(H,t)).

Note that both sensors present different behaviour: in C1 the output frequency increases with the temperature, whereas exactly the opposite occurs with the C2 sensor. The relative frequency deviations of approximately 4.5% drift per 10°C, in the neighbourhood of zero are significant when the intended uses are taken into consideration, especially with in the case of low pressures.

The individual frequency-temperature-pressure calibration of the sensors is therefore a vital phase in the quality-control of SPI-transmitted measurements.

5.1.3 Housing and flow of SPI operations

The pressure and temperature sensors and the processor card are housed in a PVC tube measuring 70 mm in diameter and 300 mm in height. The probe is connected to the SET central processor by means of an ovoid cable measuring 16 x 8 mm comprising a reinforced atmospheric pressure setting tube and a reinforced electric cable. When equipped with a 12 metre cable, the probe weighs 6 kilogrammes.

The SPI is powered by the SET at 12 volts DC. Consumption rates are 13 mA at rest and 17 mA when measuring and transmitting.

When the SPI microprocessor receives a request for a message from the SET logic, it triggers the counting procedure of the frequencies on the low sensor, on the upper self-calibration reference, on the lower reference and on the temperature sensor. In the 2-pressure sensor SP12 version (see section 5.1.4 - SP12), this procedure is repeated for the frequencies from the 'upper' sensor and from the 'upper' temperature (Figure 3).

All decoding, temperature correction and scaling of units (i.e. water level in cm, temperature in tenths of a degree Celsius) procedures are then carried out and a message is prepared. It is transmitted to the central processor via a synchronous communications link, insulated by an opto-electronic coupler, at the rate of 100 bps.

Once the processing has taken place the final precision of level measurements exceeds 0.1% of full scale ($\pm$ 1 cm). Message preparation time is approximately seven seconds and the maximum speed for tracking of levels is 0.5 cm per second.

The message is binary coded and includes 45 bits formatted as follows (SPI 1 sensor):

SPI serial number
16 bits
Water level
10 bits
Temperature
11 bits
Longitudinal parity (BCC)
8 bits.

5.1.4 Density measurement

In light of the sensitivity performances of the piezo-resistive sensors, it was worthwhile measuring the density of the fluid by differential pressure between the two sensors separated by a known veritical distance. This is the SP12 configuration. A sensor is placed at the upper and lower extremities of the cylinder acting as the probe.

This adaptation enabled us to both correct the water levels by associating pressure and density, and to transmit and memorize the variations of the density itself which, according to the environment, may be linked to salinity or turbidity. Vast areas of application are thus opened to this probe (oceanography, solid transports, sanitation, etc.).

Assuming P1 is the pressure on the upper sensor, P2 the pressure on the lower sensor, and D the vertical distance separating the two sensors, then the density ρ is expressed as follows:

$$\rho = \frac{P2 - P1}{gD}$$

This expression shows that three conditions must be necessarily combined before the density can be measured:

(a) Distance D must be as great as possible in order to obtain a large P2 - P1 with respect to the precision with which P2 and P1 are known. If the sensors' sensitivity is taken into account, the density precision would be 1% for D = 30 cm, but 0.2% for an SPI measuring one metre in height.

(b) Quite apart from the sensor sensitivity, great care must be given to the sensor accuracy and, consequently, to the calibration accuracy if a two-sensor SPI is to be used. An accuracy error equivalent to 0.4 cms water will not greatly influence a one-sensor SPI which gives the water level to the nearest centimetre. However, it will induce an error of 13 per 1000 in an SP12 measuring 30 cm high - quite a considerable error when one considers the range of densities to be measured.

(c) The probe must be placed in a vertical position so that the altitude difference between the two sensors is indeed equal to D. If the current is strong and there is a great deal of turbidity, this may entail mounting the probe on a rig.

The density is transmitted to the SET in binary using 11 bits, interposed between the water level and the temperature. The SPI2 therefore generates 56 bit messages.

5.2 Conductivity measurement

In the prototype series, condictivity was measured by a PONSELLE device, comprising a CTS10 probe with 4 moulded electrodes and an analog electronic card within the SET. The voltage output calibration, established by the manufacturer, ranges from 0 to 55,000 micro-siemens, for a published precision of 1%.

6. THE SET RECORDER TRANSMITTER SYSTEM

The SET (Système Enrégistreur Transmetteur) is housed in a watertight plastic box measuring 350 x 250 x 150 mm and weighs approximately 5 kg.

The functions of this unit are: interrogation of the SPI sensor, and collection of the information transmitted back by the sensor and by the conductivity probe. The SET also ensures that storage takes place on the CE64 cartridge and that a message is prepared in a format that is compatible with the ARGOS satellite data-collection system.

The SET has an absolute chronological basis of one year to the minute, set by adjustment buttons and displayed on a liquid crystal display. A synchronization button sets the recorder's time basis and logic.

The user accesses two 2-digit wheels which correspond to the interval (ΔT) of the SPI sensor scan, expressed in minutes (1 to 99 minutes), and to the significant water level ΔH threshold for the location, expressed in centimetres (0 to 99 centimetres). The water level is only stored on the EPROM cartridge if the variation measured between two scans exceeds ΔH. This conditional recording of the water level is called an 'intermediary message'. The cartridge also includes so-called 'half-hour' messages, the storage of which is obligatory, on the hour and on the half-hour containing in addition to the water level, conductivity, temperature and density. Some 'half-hour' message parameters may be omitted: a rocker switch on the face of the device enables the user to choose between recording or ignoring conductivity, temperature and density.

Once the time set has been switched on, the selection and display of ΔH and ΔT have taken place, the parameters to be measured have been selected and a fresh cartridge has been inserted, the normal procedure continues by initializing the cartridge, triggered by a command from the operator (push-button).

After having checked that the cartridge is empty, the system copies the characteristics chosen by the operator onto it for this particular data collection, i.e.:

Parameter	Code	Length (bytes)
SET number	Hexadecimal	4
SPI number	"	4
Parameter selection	"	2
Height sensitivity threshold (ΔH)	BCD	2
Sensor interrogation interval (ΔT)	"	2
Start-up date		
Year	"	2
Month	"	2
Day	"	2
Hour/minute	"	4

If the procedure fails, an error message is displayed on the screen which means 'cartridge not fresh' or 'write impossible'.

The SET has push-buttons for activating the SPI and the conductivity probe and for instantaneously displaying the measured values.

A command enables the alignment of the level-pressure of the sensor with the staff gauges of the site (setting the system zero).

Finally, the most recent memory address of the cartridge can be displayed at any time and its contents can also be scrolled across the display.

An ARGOS message is generated every 30 minutes, which enables the transmission of all parameters collected during the preceding two-and-a-half hours (five sensor-blocks of 6 bytes). Two bytes remain available in the message for transmission of the address of the most recent data in the CE64 cartridge.

The SET voltage is 12 volts DC (11.5 to 15 V) and consumption is 18 mA when recording is not taking place and 400 mA when the cartridge is written to. The prototype unit tested in French Guyana was fed by an 8 Watt solar panel, powering both the CHLOE (SET + SPI) and the ARGOS (CEIS - LAMD82) transmitter.

Data are coded in BCD in the following fashion:

"Intermediate" message (three bytes)	
Minute number (0 to 59)	1 byte
Water level (0 to 999)	1.5 bytes
Alignment character ('F')	0.5 bytes
"Half-hour" message (3 to 7 bytes)	
Minute number (00 to 30)	1 byte
Water level (0 to 999)	1.5 bytes
Optional:	
Conductivity in hundredths of micro-siemens (0 to 550)	1.5 bytes
Density (0 to 99) (ρ − 1000)	1 byte
Temperature in 1/10 degree (000 to 500)	1.5 bytes
Alignment character ('F')	0.5 bytes

In late 1983 the cost of a CHLOE processor, comprising one SET and one 1-sensor SPI, was approximately US$ 2,600; and with a 2-sensor, density measuring SPI the cost was approximately US$ 3,000.

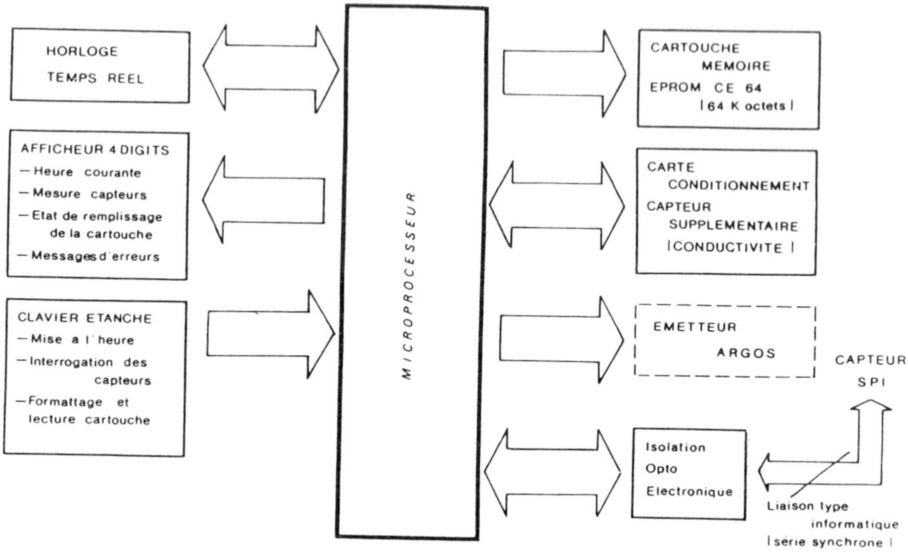

Figure 1. Chloe Processor
Block diagram of the SET recording-transmitting system

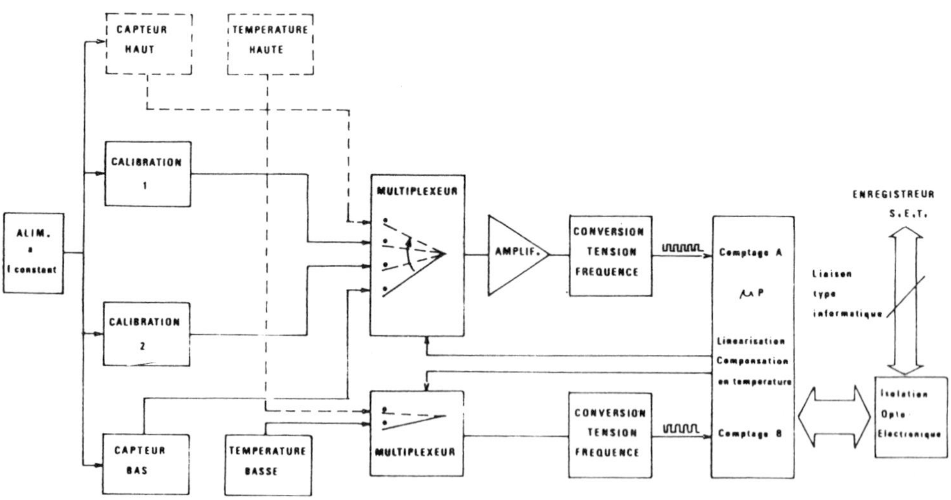

Figure 3. Block diagram of the SPI sensor

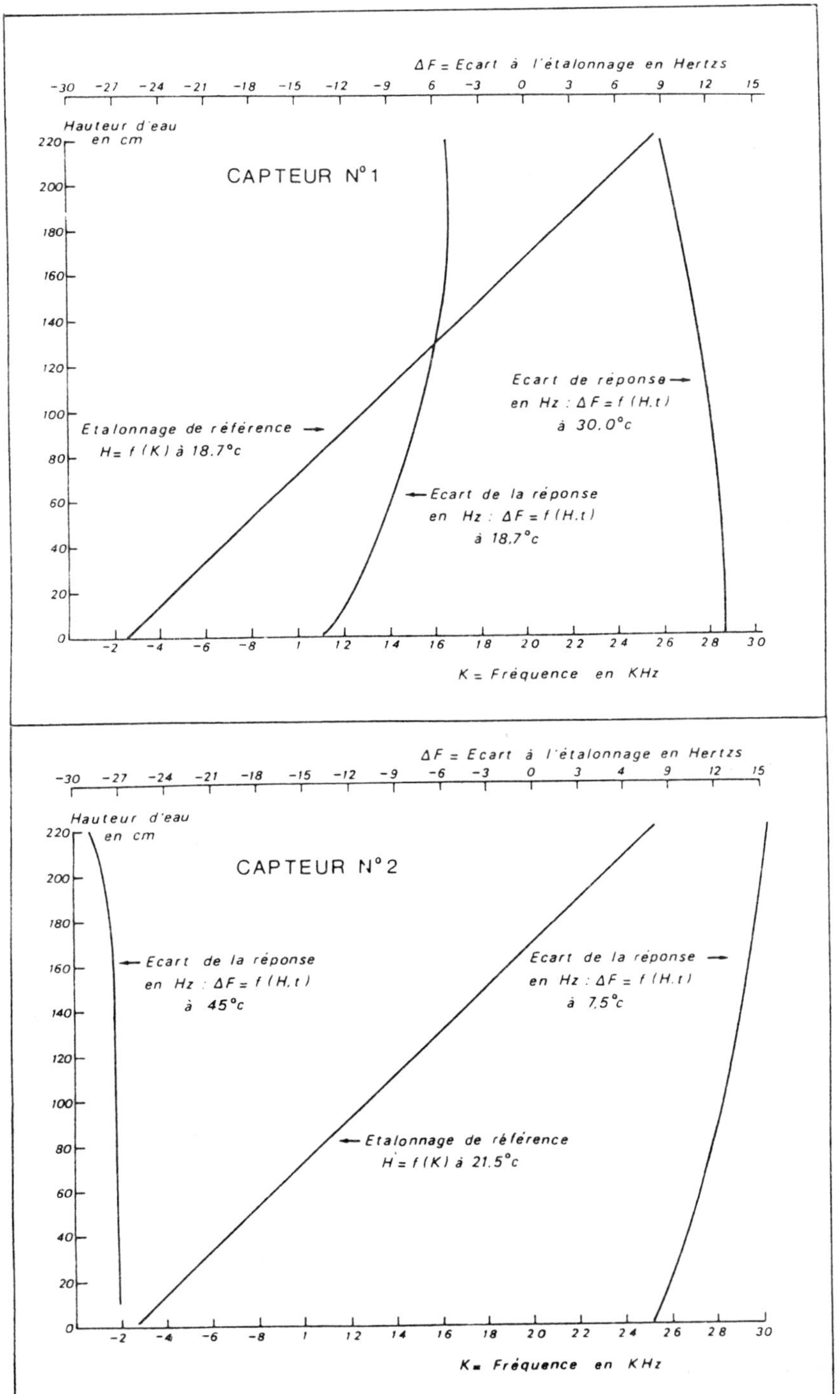

Figure 2. Pressure and temperature responses of piezo-resistive sensors

APPLICATIONS OF MICROPROCESSORS AND COMPUTERS FOR SOME HYDROLOGICAL OBSERVATIONS

L.Pálos[x] and Ö.Starosolszky[xx]

1. FEATURES OF CONVENTIONAL INSTRUMENTS IN HYDROLOGY

Up to recent times, the instruments used for observations in hydrology were characterized by the following features:

- Special design for particular functions and manufactured in small series;
- Sensors operating on a mechanical principle and in the majority of cases of fully mechanical construction, normally clockwork-driven recording device tracing a line on chart paper;
- The instruments manufactured for particular functions fail to fit into an integrated system, accurate synchronization is virtually impossible due to the clockwork mechanism; where synchronization is essential 1 to 4 pens draw traces on the chart forwarded by the same clockwork;
- The data registered are raw, meaning that - perhaps except for the discharge recorders, which convert stages into discharges - all data registered directly are recorded and stored;
- Observations are as a rule continuous, except for some maximum/minimum instruments, but with these the time of observation is normally not recorded;
- Telemetry accomplished by an encoder, or potentiometer connected to the existing mechanism, with usually an analog recorder connected at the receiving or processing station.

A review of the waterlevel recorders available a few years ago on the market will clearly illustrate the situation. Each type and model of each manufacturer operated with floats and counterweights of different size, but the dimensions of the recording devices, the size of the chart tapes, the design of the clockworks and winding mechanisms were also different. This diversity could be attributed to some abundance of assortment, which was, however, not so rich by far, when viewed from the side of the user. The advanced mechanical recorder was made applicable under a variety of conditions by the variable amplitude- and time scales /tape speed/, in contrast to the earlier situation, when the choice of model had to be made in accordance with the intended application /fixed amplitude- and time scales/.

The instruments required no special skills from the operators, they were adjusted to the availability of manpower, but needed actually rather frequent attendance, such as checking, chart replacement, ink refilling, winding the clockwork, etc.

The spreading use of computers for data processing and forecasting work has raised in the late sixties and early seventies the need for digital, machine readable data recording readily accessible even to computer terminals. In response to this demand computer technology created the required facilities. Mechanically operated punched-tape recorders have been developed and manufactured /Fischer-Porter, Ott/ for combination with the mechanical instruments. Punched tapes were suited to computerized processing, but these devices required higher and service-repair facilities, skills to operate than the clockwork-driven recorders. The time of sampling could be varied within certain

[x]Head of section for instrumentation; [xx] Director
Institute for Hydraulic Engineering, Research Centre for Water Resources Development, VITUKI, Budapest, Hungary

World Meteorological Organization (ed.), Microprocessors in Operational Hydrology, 39-47.

limits,and with some instruments checking was facilitated by a small recording mechanism /thus two types of data registering functioned parallel to each other at the site/.

The recording devices adhering to the mechanical principles were developed to the highest attainable level, but the problem of integration into a consistent system remained still unanswered. The introduction of electronic devices became thus unavoidable. Integration into a system has brought about fundamental changes in the electronics built into the mechanical devices, made possible by the advent and mass production of microprocessors in the early seventies. The first radical changes were of but experimental character, nevertheless they forecast the possibility of replacing completely the mechanical devices by the eighties.

2. THE ADVANTAGES OF MICROELECTRONICS

Extensive development in microelectronics /microprocessors, pocket calculators, chips/ has opened the avenue for the development of the next generation of hydrological instruments, the work on which has already been started to incorporate the following features:

a. Introduction of new sensors, which produce an electrical signal, are less sensitive to mechanical effects, such as silting, clogging, coupled with the use of pressure transducers instead of the wave-responsive float sensors;

b. Variability of the time of sampling with provisions for automatic compliance with the sampling instructions at the site;

c. Low-power transmission or amplification of the signal;

d. Compact storage of the signal in digital form at the site;

e. Primary processing of the raw signals according to different requirements /storage of time-averages, extreme values alone, calculation of means, standard deviation, smoothing/;

f. Relatively simple communication of the data observed by remote control or telemetry;

g. Automatic error warning;

h. Universal output signal for all elements registered;

i. Use of components and even instrument units /e.g. recorders/, which are mass produced and thus cheap;

j. Complete elimination of moving components at the observing stations.

With the advent of the Complementary Metal Oxide Semiconductors the power consumption of integrated circuits decreased in operation from mA to uA orders, while in stand-by mode from uA to pA orders, creating thus the possibility of developing instruments with broad potential applications, comparable to those of the clockwork devices. The microprocessor, the ramified functions of which can be controlled by programming, and the programmeable complex circuits incorporated into a single silicon chip have made the development of versatile general-purpose digital devices possible. At the same time, programming gives the unit the required personality.

Microprocessors have become available in ever increasing varieties and the devices using them are intended for mass production. The calculator or microcomputer chips appearing simultaneously on the market can be used in hydrological instruments, in some cases without programming which would otherwise be necessary if microprocessors were used.

Development proceeds on a world-wide scale, and some similar applications developed successfully at the Research Centre for Water Resources Development will be reported on subsequently.

3. THE USE OF POCKET- AND DESK-TOP CALCULATORS IN DATA COLLECTION

Pulse counting devices connected to calculators find their most effective applications in combination with revolving current meters, since in addition to counting the revolutions, they are suited to provide direct readings and printouts of flow velocity and even discharge.

In the arrangement shown in Fig.1 the programmeable calculator is connected to 5-decade counters, to which sensors with pulse outputs can be coupled directly. The device is capable of operating in two modes. In one of these the pulses are counted over a fixed period of time, whereafter the result is entered electronically into the calculator. In the other mode one or several counters are used for timing, thus sacrificing some of the channels available, but enabling us to cover a very broad time span.

The advantage of the device shown in Fig.1 is that the need of microprogramming is eliminated, the instruction set of the calculator adopted being available to the user. For measurements with a micro current meter /MIMOSA/, the device was developed jointly with the Delft Hydraulics Laboratory in 1975.

The device has been used extensively also for radiometric soil density and moisture measurements. Units sensing neutron and gamma pulses have been connected to the counters, the calibration equation has been entered into the programme, so that the densities and moisture contents could be displayed or printed out directly.

The use of compact programmeable calculators has the advantage over microprocessors that a smaller number of components need be purchased for constructing the device, since the display and keyboard, the printer and the magnetic card entry are available in a single unit.

In these devices no separate display has been used for the individual counters, since the display of the calculator served this purpose. Data entry has been accomplished by modifying the calculator slightly in order to mount a connector for the matrix points of the keyboard.

One of the devices has been developed for flow measurements from cableways. Int this case two counters count both ways and store the vertical- and horizontal position of the current meter, while other two counters register the pulses and the time.

Parallel keyboard entry is an advantage in using pocket calculators as instrument components in cases, where only a few devices are to be constructed and the problems to be solved requires sampling at relatively low frequencies /10s/.

Larger measuring systems can be composed in a structure corresponding to the recommendations of the IEC 625 and IEEE 488. The Institute has produced measurement-oriented peripheries, 10 channel pulse counter, 10 and 16 channel A/D converters, current meter calibration control equipment, etc., for connection to the EMG 666 type computer manufactured in Hungary. In this case the computer controls on the one hand the frequency of sampling /the interval between observations/, on the other hand, it collects, stores and performs primary processing of the data observed, e.g. the waterlevels in a row of groundwater observation wells, the waterlevels observed at several sluices on a canal and pulses of the current meter. Moreover, it can be used to control the sluice gates automatically. The computer manufacturer produces further devices of universal applicability, into which special electronic units can be mounted. This allows us to utilize the advantages of mass production in manufacturing devices for particular applications.

In the sphere of applications corresponding to the recommendations of the IEEE 488, low power consumption devices cannot be incorporated at all, or only at important compromises in consumption. This fact and the low price of pocket calculators having no I/O facility warrant the continued application of parallel data entry by means of the keyboard matrix according to Fig.1.

For combination with the Model HP 41C pocket calculator having an I/O interface, devices have been developed, which enter the digital output signal of the counters or the A/D converter through the WAND /BAR CODE reading pencil/ of the calculator. This device is used to measure ice temperatures simultaneously at several points in a profile of the ice cover. From the viewpoint of design it is of interest to note that fluctuations in the 0 level of the measuring amplifier have been eliminated by means of the software, in that a temperature-invariant manganine resistor has been connected to one of the measuring channels instead of the platinum thermometer. The same solution has been adopted in combination with the sensor of the chemical /salt dilution/ flow measuring device, so that after entering the basic data, such as initial conductivity, water temperature, the pocket calculator yields directly the flow rate desired.

The development of the interface system for compiling low-consumption HP-IL measuring sets appeared thereafter logical at VITUKI. By this technical solution both field- and laboratory devices could be developed under conditions identical with the degree of freedom corresponding to the application technique corresponding to the recommendations of IEEE 488. The first HP-IL interfaces were acquired in January, 1983 and by now IL interface compatible 8-channel A/D converters and a 4-channel special device with up-down counters are available for cable-way streamflow measurements.

4. APPLICATIONS OF MICROPROCESSORS

The applications of microprocessors-based devices will only be outlined below. The manufactured microcomputer board-comprising the M 6800 processor has been used as a transient recorder and as a special calibration device, as the RAM element peripheral of the EMG 666 Model computer. For observing the waterlevel in wells a low-consumption data collector has been manufactured starting from the M 146805 CMOS processor. Capacitance probes, or piezoelectric crystals can be used as the sensor element. The battery-powered device is suited for one-year unattended operation, and data readout is provided in the HP-IL compatible system. The CMOSRAM memory serves for data storage, with a separate lithium battery securing data conservation. With this device no practical experiences are available thus far. Data collection involving a semi-conductor memory has been carried out under field conditions using an HP-41C calculator. The aim in developing the micro-processor device is to make this solution cheaper, and to arrange it in a mechanical construction suited to be installed directly into the observations wells, providing thus protection against unauthorized interference and temperature fluctuations.

5. EXPERIENCES OF APPLICATIONS

Since microelectronics gave new tools for hidrological measurements and the improvement enables users to collect hydrological data on a more rational bases, it is interesting to note the experiences of the operational applications. In this respect two examples are introduced, namely the continuous measurement of ice/water/air/bottom temperature and the current metering by a cable way.

Since ice temperature is the major factor in calculating and predicting ice pressure and ice loading capacity its observation is necessary in different depths of ice cover in lakes. Continuous observations on the Lake Balaton, at Siófok were carried out by platinum thermometers at the bottom of the lake, in the depths of 100, 60, 40, 20 cm, at the surface of the ice cover and in the air. The temperature was sampled three times consecutively in each point at every hour and was averaged at each fixed observation period. Sampling was thus immediately checked, the value of the three, almost simultaneous observations should be within a limited range. The observed values and their deviations were printed out by the HP 41C pocket calculator interface. In order to check the original data their values were also printed at this pilot application. The recorded data were plotted/Fig.2./

As it was expected the highest fluctuation was observed in the air temperature, but still slight fluctuations in the water temperature and in the deeper layers of ice have been measured. The temperature of the surface of ice cover follows the air temperature variations with a small time-lag and with a certain reduction.

The adjustable observation frequency, and the almost simultaneous multisampling were of great advantage of the observation system, and the ice temperature records demonstrated the applicability of the pocket calculator.

On medium size rivers, in particular, during floods, the discharge measurement by cable-ways is a common method.

The electronic system is basically the same as it was introduced earlier on Fig.1. The only difference is that the impulse counters are "up-down" counters.

The system follows automatically the operation of a cable-way current metering station. The positioning of the current meter is performed by suspension cables. On each cable winch a reel, as a distance meter was adjusted to the existing mechanical counter. The distance is measured by special encoder logic consisting two micro switches. A special encoder logic regarding to angle direction of the reels gives the up - or down control impulse for the counter. The horizontal distance is positioned to "0"at the bank, manually while the vertical distance/depth/ automatically when the current meter reaches the surface. Suspending the current meter the impulses of the revolutions are measured by a revolution counter. The measuring conditions can be set either by the required number of impulses or by the required sampling interval. The calculator has a flexible, adjustable programme. The calculator can be programmed to make a print-out at the end of each measuring period or at any desired interval.

Reaching the bottom the specific discharge of the vertical can be calculated and printed out, or put into the memory. Calculating consecutively the specific discharges of the selected verticals of constant distances their summation gives the measured discharge. The programme should contain also the rating curve of the current meter used at the specific case. At this measurement perpendicular flow to the cross section was assumed.

The cable-way installed on River Rába /a tributary of the Danube/ is shown on Fig.3, and the calculator based measuring system on Fig.4.

The experiences of the application may support the idea that micro-calculator based current metering promotes accuracy, reliability and reduces malfunctioning of the measuring system. The continuous print-out by a printer is strongly required due to the ellimation of certain unexpected, unprogrammable effects during the measurement. It is recommended to use liquid cristal displays due to invisibility of light emitting diode /LED/ signals on day-light.

6. CONCLUSIONS

The applications outlined briefly in the foregoing demonstrate that microprocessors and pocket calculators open new opportunities in the domain of hydrological observations for

- sampling at intervals adjustable without limitation,
- adjusting the duration of sampling,
- averaging the samples taken,
- storage and display of the values desired,
- primary processing of data at the site /station/ and thus
- merging the processes of observation and processing.

A process has thus been initiated, which may bring about radical and rapid changes in the hydrological data acquisition system and in the hydrological observation network.

BIBLIOGRAPHY

Starosolszky,Ö.: Trends in the development of hydrometry. /IASH-WMO Symposium on Hydrometry, Exeter, 1982/ Hydrological Sciences Journal, 28. 1,3/1983 p. 103-123.

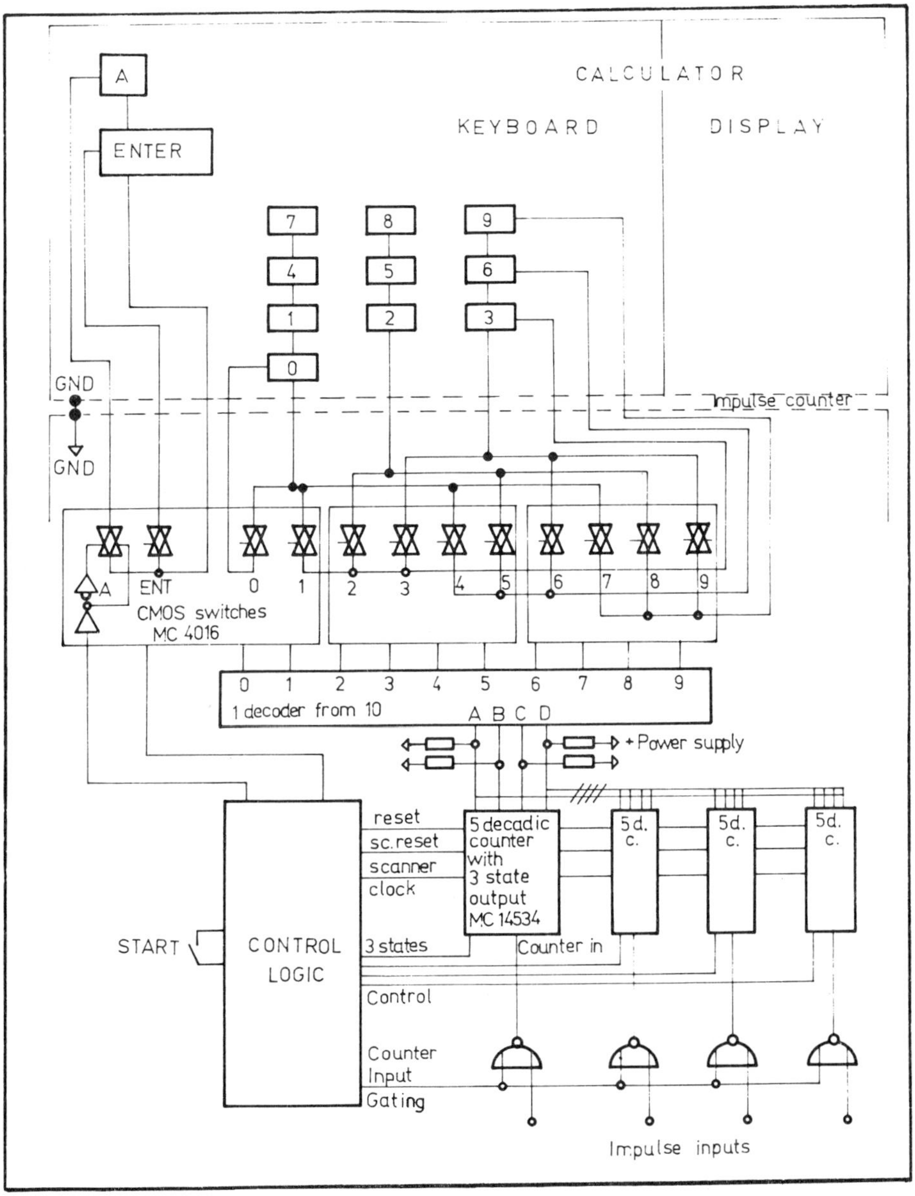

Figure 1
Calculator operating parallel with the keyboard

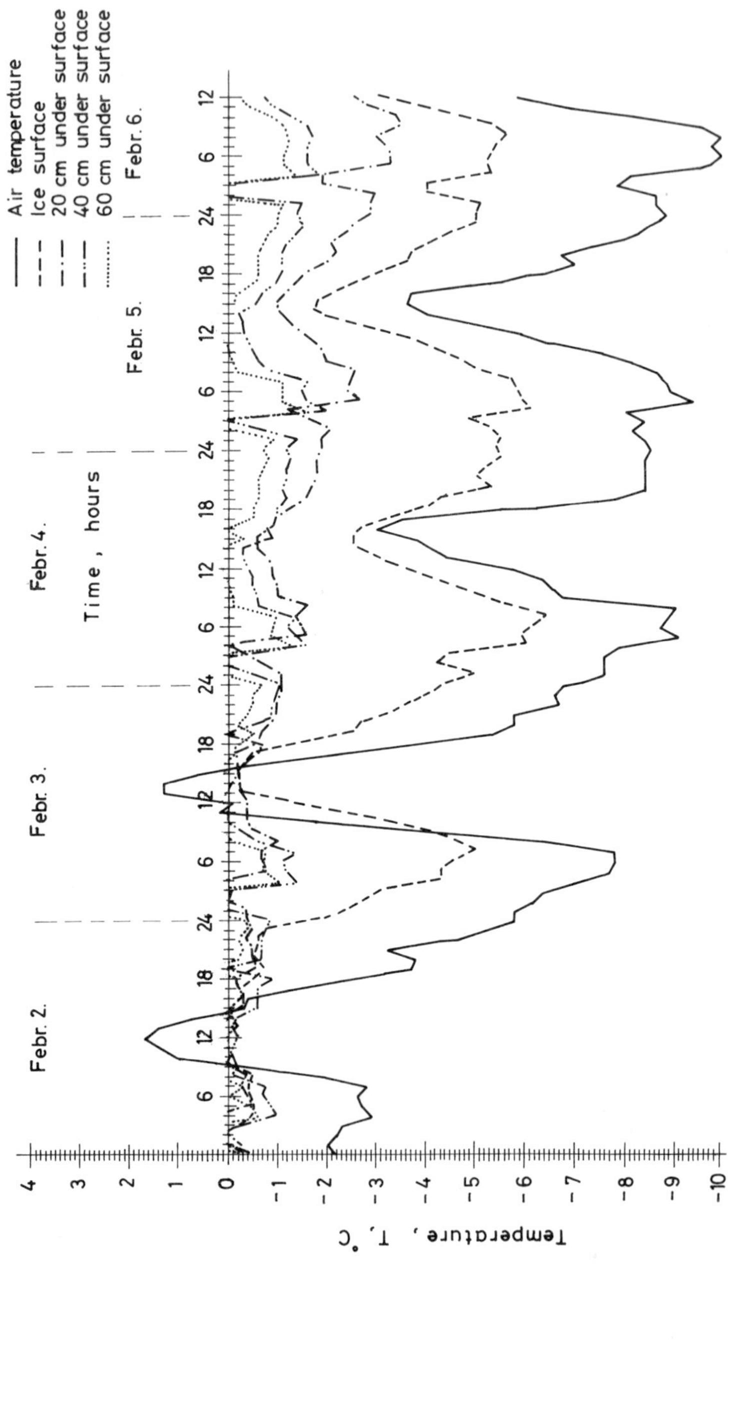
Air temperature
Ice surface
20 cm under surface
40 cm under surface
60 cm under surface
Febr. 2.
Febr. 3.
Febr. 4.
Febr. 5.
Febr. 6.
Time , hours
Temperature , T, °C

Figure 2

Figure 3. Cableway installation, River Rába

Figure 4. Calculator based mesuring system

RELIABILITY OF MICROPROCESSOR-BASED FLOOD-WARNING SYSTEMS; CHOICE OF ARCHITECTURE AND DEGRADED MODE PROCEDURES; FRENCH EXPERIENCE

by P.A. Roche*

1. INTRODUCTION

The French Government has launched a large-scale modernization programme for its flood-warning systems with three main thrusts: automation of real-time data collection, updating and improvement of flood forecasting methods, and reorganization of the public information system. In this paper, we shall discuss in more detail how, and within what limits, the introduction of the microprocessor and microcomputer in the data-collection system for flood warnings has increased the reliability of the whole sequence of operations.

2. SECURITY: A FUNDAMENTAL PROBLEM

One could say without exaggeration that, in a data transmission system, it is the exceptional situations which have to be taken into account and the need to guarantee performance in such cases which ultimately define the system even in the smallest details.

By leaving less room for human intervention, and the latter's failings (although also its capacities for adaptation), automation brings both advantages and risks as regards security. The circumstances in which the system is particularly useful are generally trying ones since floods may cause a breakdown in the electrical power supply, broken cables, wires and deterioration of measurement equipment. A system with low reliability is useless since it will not operate when needed.

Such cases are not rare. In July 1977, the floods on the Gers and other tributaries of the Garonne carried away telephone lines and the alarm could only be raised by the radio links of the Gendarmarie Nationale. In 1978, the submersion of the cable linking Dakar airport with the receiving station for the satellite data transmission system shortcircuited the wires and stopped the routing of messages for several days.(1)

It seems important to have standby arrangements available so that the system can function in normal mode, but it also in a certain number of degraded modes which preserve as far as possible its more important functions.

---

* Civil Engineer at the Sub-Directorate of Inland Waters, State Secretariat for the Environment and Quality of Life

(1) Quoted by J. Callede, Teletransmission requirements of the Financial Agency of the Seine-Normandy Basin, ORSTOM, 1980

World Meteorological Organization (ed.), Microprocessors in Operational Hydrology, 49-57.

3. POSSIBILITIES OF EXTERNAL BACK-UP

In order to increase the security of several independent systems, we are seeking methods for them to back each other up.

Several flood-warning systems in the lower Loire Basin use the same transmission vector and compatible hardware. For negligible extra cost, security has been improved considerably by installing in each collection centre the necessary programs for interrogating stations managed by other centres. As a last resort, one should have a list of people to contact in each commune concerned who are capable of going to take a reading on the staff gauge. To do this, the person must know where the staff is and how to take the reading. This also means that, as back-up, there must be a graphical method of forecasting and that the forecasters know how to use it. In general, manual methods are being kept at every level of the system. The transition from automatic to manual mode should be carefully studied.

4. THE STRUCTURE AND OVERALL RELIABILITY OF THE SYSTEM

The simplest way of improving a system's reliability is to increase redundancy. If the same task can be carried out by several pieces of equipment and if there is a very reliable procedure for switching between them, the probability of outages of the whole system is theoretically very low.

Excessive complexity in the structure nevertheless generates breakdowns, in itself. The switches are relatively fragile parts and devices are frequently used in which the degraded mode is not strictly automatic but includes manual operations.

But, beyond doubling the equipment for carrying out a particular task, the aim has been to diversify the equipment in order to have back-up means which would not be sensitive to the same environmental disturbances as the means for normal operation. Ultimately, it is the diversity of the means for transmitting the information which determines the network's structure.

Several network structures can be envisaged:

- Network with a single collection centre (Seine basin) - This collection centre is the hub of the whole network and everything depends on its operation. This is the case with small networks or when secondary centres are merely peripherals of the main centre. the secondary centres are not then linked to each other. The links with the stations may be more or less redundant.

- Network with several collection centres backed up by a main centre (Saône basin (in the project stage)) - For greater security, a main centre can still carry out back-up collection, but the secondary centres collect the information with which they are concerned, each for its own needs. In general, the main centre is that of the intermediate basin, which is situated in such a way that it can back-up both upstream and downstream centres. The secondary centres are linked only to the main centre.

- Network with several collection centres capable of backing each other up (Loire basin) - The most reliable solution is that in which the whole group of centres constitutes together a network in which each can be backed up by one, two or more of the other centres and in which each performs its collection autonomously.

Certain transmission vectors may be reserved for the main centre which, thanks to the duplication of the links, will then serve as ultimate back-up for the secondary centres, but in normal operation it will control the overall management of the centres. In some cases, this function of the main centre disappears. We then have again the case of autonomous systems backing each other up.

The structure of the system cannot be properly studied without knowledge of the transmission means used. Indeed, although surface radio telecommunications provide for a good redundancy in the communications systems, it also leads to a certain centralization of the processing hardware, which then becomes vital. In the case of the telephone network, on the other hand, the number of centres participating in the collection may be extended without hampering the system's operation. Mixed solutions are thus of considerable value for reliability.

5. RELIABILITY OF EACH OF THE SYSTEM'S ELEMENTS

The security provided by the structures mentioned above depends greatly on the operational reliability of each of their elements.

By examining of the system's structure, one can define those elements whose operation is essential. An element must then be selected whose reliability is almost perfect. If this is impossible because of technical constraints, the structure is poorly adapted to the situation and this element should be duplicated. However, the elements remain key areas which must be studied in greater detail.

5.1 Power supply

The operation of each element requires a stable power supply which is either permanent or available at any time.

This power supply can be rendered secure by various devices:

- Power pack (back-up power supply) for centres and/or for relays;
- Buffer battery (accumulator) for measurement stations and/or for relays;
- Protection against lightning.

5.2 Central sites

A breakdown in the central computer is more or less serious, depending on the structure. Data-processing equipment is subjected to permanent operation, unlike the micro-computers used by the public at large. Whereas, in normal use, a micro-computer can easily withstand small mishaps (bad initialization, inadvertent stopping of a procedure, accidental erasure, etc.), real-time use requires a "faultless" run when the operator is absent. In particular, since this equipment is quite sensitive to momentary power failures, the power supply must be stabilized. It is possible and preferable to use hardware and software which offer safeguards in the event of outages and easily permit manual re-initialization of the system. But it is still necessary to use procedures to alert the managers in case of an outage of the central computer. Of course, this function must be undertaken by special equipment with excellent reliability.

Commercial micro-computers are, however, available at very low prices. Their maintenance is also cheap. When performances are sufficient, the use of two micro-computers in parallel, with switching governed by a very simple industrial processor, ensures almost total reliability. For seven years, the same micro-computer has been managing the Garonne network (120 stations, 16 relays and 4 sub-centres) permanently and without any noteworthy incidents. The only breakdowns observed were due to untimely intervention, when the equipment was under power. Nevertheless, this micro-computer has undergone several maintenance services. The mini-computer and peripherals for the network of the River Gard require about 4 services per annum.

In these two cases, where single processors are used, the maintenance focusses particularly on the hard disk which is necessary for storing the many programs.

Although the choice between micro-computers and mini-computers loses its importance with the developments foreseeable for this hardware, the questions raised concerning the organization of the centre will still remain for a long time: should each computer be strictly devoted to a particular task, or should an architecture be found in which they play more symmetrical roles?

The arrangements now being adopted for the new stations entail a certain amount of specialization, whereby the equipment is devoted, on the one hand, to data collection and management, and, on the other, to digital processing and storage. In the case of the Dordogne network, which is being modernized at present, it is planned for the collection to be made by duplicate equipment which will also be responsible for alarms (i.e. the highly reliaible part). On the other hand, the calculation and archiving will be done on a mini computer, for which maintenance delays in the order of 24 hours will be acceptable.

Actually, it is the central equipment which has the greatest requirements for fast diagnostics. As soon as the structure of this central equipment becomes complicated (around a single processor), we use the diversity of "intelligent" hardware so that each element can be given external tests by another. A file of the various operations made by each element can thus be consulted on another, which continues to operate (case of two microcomputers in parallel, for example).

The mechanical peripherals are always fragile (printer, but especially diskette reader, plotting table, etc.). One must therefore ensure that they are not essential to the operation (print-out of all of the keystrokes made at the console, frequent back-up on diskette, etc.).

The non-essential peripherals may be given simple maintenance. The others must be capable of immediate replacement.

5.3 Transmissions

Many external events can damage transmissions either temporarily (disturbance) or until repairs are carried out (degradation). For satellites, we can ignore this vulnerability, since the reliability problems are more internal to the satellite.

For radio links with ground relays, the most important factors are disturbances (interference from other transmissions or magnetic storms, propagation difficulties, etc.). However, these disturbances are very weak when the radio fields exceed 20 dB.

Transmission faults are very rare in normal meteorological conditions, but may reach 2 %(1) during disturbed periods. For all surface-based equipment, the risks of destruction are linked with lightning(2).

The telephone network in France has considerable error rates. However, in practise, the call decoder is capable of detecting them and all erroneous messages lead to a repetition of the call(3).

(1) Estimation by A. Leoussof for the Gard network (15 stations and 2 relays in service since three years ago).

(2) Total protection against this risk does not exist. In the Gard network, the stations are very carefully insulated: photovoltaic cells are used for the power supply (no connection to the mains supply), and good quality earthing is used (impedance of less that 5 ohms). During a thunderstorm on 30 August 1982, two stations were nevertheless hit.

(3) In the case of the telephone, the reason for increased collection delays is the high number of calls which are not connected during periods of network saturation.

In fact, beyond the physical disturbances which lead to complete interruptions in communications, the numerous tests and calling procedures now make it possible to obtain transmissions of excellent overall quality. The advantage of programmable systems can be seen here, as they improve the tests and procedures on the basis of experience. When this has not been the case, considerable difficulties arise. On old hardware, the coding and decoding of information was carried out by hard-wired logical circuits. In the case of the Dordogne network, for example, it has been observed that certain messages had a systematic decoding error which rendered the data unusable (loss of the last digit of the value transmitted). While such errors are easy to correct on a system using microprocessors (program correction), the hard-wired circuits require the intervention of electronic specialists.

5.4 Observing stations

For the most part, observing stations are links in the chain which have mechanical parts and are subjected to particularly difficult environmental conditions (temperature, humidity).

In automated systems, it is the on-site sensors and graphical recorders which cause the most frequent breakdowns. During their three years of operation, the fifteen stations in the Gard network have required, on average, six interventions per year, per station.

With real-time operation it is possible to detect anomalies quickly (blocked raingauges, broken or jammed windvanes, jammed float gauges, out-of-order pressure sensor river level gauges, and faults in the closed-loop control systems).

A certain number of these failures can be detected locally, if the station has independent logic (failures in the power supply, breakages, faulty connection with a sensor or a coder). The advantage of this is that a diagnosis can be made of the validity of the data transmitted even before going to the site and repairing the equipment.

Although the need for good-quality information is obvious in real time, one should also emphasize the need for critical assessment and analysis in deferred time. Telemetered data are insufficient for this purpose because of the various disturbances which may occur en route.

Local on-site storage of information is therefore always considered appropriate even on those networks on which the information is reported in real time.

Two time scales can be distinguished for this storage. In the short term, the aim is above all to avoid any loss of information following a link outage. In this case, volatile memories (RAMs) can be used. In the long term, the need is for a recording which can be compared with the transmitted data. In such cases RAMs supplied by internal batteries or non-volatile static storage (ROM, PROM or EPROM) should then be used. Electronic systems have certain advantages in terms of reliability as compared with the conventional paper charts, owing to the absence of mechanical parts[1].

(1) Many attempts have been made to use magnetic tape cassettes (as used in the Electricité de France - Division Technique Générale). Here we are speaking about static storage media which seem to have greater prospects for the future in view of their recording capacity and resistance to high temperatures.

6. MANAGEMENT SECURITY: SAFEGUARDING PROCEDURE

The design of the system's security does not stop with the detection of breakdowns, evaluation of their probability and of their impact. One should also envisage safeguarding procedures. The fact that a system is driven by microprocessors at each level permits these safeguards to be created in the best conditions. The hardware should be capable of restarting after any breakdown with the least possible damage; this is what we shall call management security. In the case of a relatively large catchment a breakdown which is repaired quickly does not endanger the system's operation. However, the necessary data must still have been stored on stand-by at the various levels of the acquisition sequence. In the event of a breakdown in the central computer, nothing prevents the observing stations from continuing their operation, but they must then be able to store the information collected locally for the duration of the repair work; their memorized data should also be capable of being retrieved at the centre when they restart (see above). In these conditions, the delay may be quickly made good and the quality of the service is not lowered.

7. SOFTWARE FOR FLOOD FORECASTING

A quick examination of the reliability of data collection systems has shown that, even with modern hardware, one can scarcely do more than manage the disturbed periods as well as possible, but certainly not take precautionary steps against all real-time failures. For this reason, perhaps contrary to expectations, the use of a computer for flood forecasting does not have the sole advantage of the possibility of running very complex hydrological models using very many variables, but there is also the possibility of carrying out very many tests and multiple consistency checks, managing access to prior flood simulations, etc. Indeed, the operational objective is not so much to improve forecasts which are already good when all of the information is available, as to ensure this quality, which is sufficient to set off alarms correctly, in all circumstances, and to avoid the gross errors which may arise through the automatic application of a particular calculation procedure to insufficiently scrutinized data.

Flood forecasting software is being developed to take into account these real-time difficulties.

We shall now discuss separately the tools for display, calculations to assist decision taking in real time, and flood simulators.

7.1 Display

Graphical display is the first stage, at which the forecaster who has been alerted in the middle of the night can gain a preliminary idea of the events taking place. It is important that these representations be not only illustrations of the phenomenon, but that they can also serve as working documents for any manual method which the forecaster might use, especially in case of system failure.

7.2 Algorithms to assist decision taking

The principle of these algorithms is as follows. Not one, but several forecasting methods are simulated in real time and the consistency of the various results is evaluated in order to establish the most probable hypotheses on which the forecasts will be based. This makes it possible for the decision to take into account the uncertainty of the adjustment of the various models to historical floods, and to test, on any type of model, whether or not it is capable of supplying an acceptable interpretation of the observed data. When excessive discrepancies appear, it is then possible to use one or other of the methods which do not use all of the information. When all of the possible decisions are represented, a very large number of choices arise which are not easy to make, such as until when will it be preferable to use a complete method, even if the missing data must be evaluated by analogy with other variables or qualitative information (non-digitized images of meteorological radars, for example), or should the "graceful degradation" method be adopted, which directly uses a back-up observation station post? In cases where data are not missing but only suspect, two decisions are possible, namely (a) to consider them as erroneous and replace them by estimates, or even change the forecasting method with the risk of losing precious information; or (b) to assume they are correct, and risk making considerable errors if they are the result of some chance occurrence. Which entails the least risk? This is one of the major problems in improving quality in alerts.

There is a need to readapt real-time forecasts, explore all of the decision possibilities, evaluate their impacts, and simulate various scenarios for the evolution of the precipitation, in order to provide approximate evaluations for the longer term.

The development of such procedures is under way. They are a necessary complement to the automatic collection systems, which permit the forecaster to continue using the resources of such material, even when disturbances occur.

7.3 Flood simulators

The training and motivation of personnel providing the round-the-clock flood-warning service are very varied, and it is not always specialists who are led to take part. It is therefore extremely useful to be able to use the data collected during previous floods, or even to generate synthetic floods from a simple model, in order to put this personnel into practice situations in which they will learn how to respond to events. The equipment at the collection centre is perfectly well adapted to this training, if the functions of collection and alarm are separated from the calculation body used for this exercise, and if the trouble is taken to archive important events in a manner very similar to real-time management. The real-time operation software can then be transformed into simulators with less programming effort.

Since this practice is becoming generalized, it is now apparent that this system can also be used for trying to improve forecasting methods, and that it is being used just as much by non-specialists as by hydrologists.

8. CONCLUSION

The development of higher-speed flood-warning systems is essentially linked to the improvement of their reliability. The introduction of micro-processors to replace both electronic hardware with hard-wired logic and electro-mechanical systems has made for considerable gains in all areas: reliability, ease of maintenance, adaptability, safeguards and operational checks. The central sites are equipped both with management computers for automatic driving of the interrogation cycles, and computers which provide simulations of forecasting methods, elaborate consistency checks and numerous calculations supplying information to assist alarm decision taking, which had been, until now, totally inaccessible in real time. Today's software should be designed for managing situations of degraded mode operation, since such situations are always possible.

The main thrust will therefore now be towards improving procedures for the man-machine dialogue, in order for the forecasters be able to exploit all of the potential of the hardware installed.

The proper operation of these systems nevertheless presupposes scrupulous maintenance. The pioneer teams, who developed the systems which are currently operational, are now passing on to the newly arrived personnel, the motto which is written beside the system diagram of the Gard network: "An automatic system is not a system left alone".

A REAL-TIME RAINFALL DATA ACQUISITION SYSTEM

B.Y. Lee *

INTRODUCTION

Hong Kong lies just within the tropics and is located on the southeast coast of the Asiatic continent. It is susceptible to effects of active summer monsoon and tropical cyclones which often bring heavy rain. A number of severe rainfall occasions in the past caused extensive landslips and floods which resulted in major losses of human life and property. As a measure to mitigate damage, the Royal Observatory, Hong Kong, implemented in 1983 public advisory services on landslips and floods. Because rainfall is an important contributing factor to landslips and floods, the Royal Observatory identified a need to implement a rainfall data acquisition system (RDAS) in which data from a network of automatic rain-gauges would be relayed back to the Central Forecasting Office in real time.

This paper describes the RDAS, which was implemented in November 1983 and designed to:-

i) receive and monitor inputs from a network of 20 rain-gauges each connected to the Royal Observatory by telephone wires;
ii) transmit rainfall data accumulated over a fixed time interval to a main computer system for display and archive at the Royal Observatory; and
iii) store raw data in a computer-compatible form to facilitate subsequent analysis of short-duration rainfall.

Tipping-bucket rain-gauges are used in the system and they operate in the following manner. Rainwater is funnelled into one of the two buckets which rest on a pivot in such a way that when a given amount of rain (normally 0.5 mm) has been collected in the bucket, it tips and momentarily closes a magnetic reed switch. The tipping action discharges the water and at the same time causes the other bucket to start collection. The momentary closure of the reed switch indicates that the given amount of rain has fallen. The raw data referred to in (iii) above include the time at which a tip of the rain-gauge bucket occurs and enable an estimation of the rainfall rate under intense rain situations (Donnelly 1979).

---

* Scientific Officer, Royal Observatory Hong Kong

World Meteorological Organization (ed.), Microprocessors in Operational Hydrology, 59-68.

DESIGN OBJECTIVES

The detailed design objectives are as follows:-

(1) Signals representing rainfall collected by a tipping-bucket rain-gauge should be distinguished from random noises which might be present in telephone wires.
(2) The effect of contact-bounce which occurs in rain-gauges employing magnetic reed relays should be eliminated so as to avoid over-estimation of rainfall.
(3) Rainfall data should be pre-processed and then accumulated before being sent to the main computer of the Royal Observatory at specified time intervals so as not to overload the main computer.
(4) In order to minimize the loss of data in case of communication failure with the main computer, rainfall data accumulated over the past quarter of an hour and over the past hour, and that recorded since midnight should always be recorded on hard copies and displayed on an optional video monitor.
(5) The raw data, which include the time at which a signal (i.e. a contact closure resulting from a tip of the bucket) is received from a rain-gauge, should be recorded on a computer-compatible tape to facilitate data analysis.
(6) To safeguard against interruption caused by power failures, power back-up at the central station should allow for at least a certain period of continuous and uninterrupted operation immediately after a power stoppage. Also, to avoid the process of manually adjusting the time after a power stoppage, timing should be provided by an electronic clock which is backed up by separate batteries.
(7) To minimize damage caused by thunderstorms and power surges, lightning and transient protection should be provided at field sites and at the central station.

THE DESIGN

The system design of the rainfall data acquisition system is outlined below. Details of electronic circuitry and software flowcharts are given in a Royal Observatory Technical Note (Lee 1984).

Transmission of rainfall information

Rental of low-speed, dedicated telephone lines is relatively inexpensive in Hong Kong. The lines consist of sections of land lines and radio-frequency links and are used in the communication between the rain-gauge and the central station. In order to minimize interference from random noises present in the telephone lines, the electrical pulse generated from the rain-gauge is converted into a data stream with a pre-defined bit pattern (i.e. a series of 'lows' (Ø's) and 'highs' (1's)) before being frequency-modulated for transmission on the telephone lines. The probability of random noises having the same wave pattern as the modulated data stream diminishes with increased complexity in the bit pattern.

In the present design (Fig. 1), the field equipment consists of a rain-gauge and an interface box (measuring 27Ø mm x 135 mm x 2ØØ mm) which is connected to the telephone wires. Plate 1 shows an interface box connected to a rain-gauge. The interface box consists of a single-chip microprocessor controller (Mostek MK38P7Ø), a light-emitting-diode (LED) display to show the collected rainfall and a low-speed modem (Motorola MC14412) for modulation of the output signal. The LED display is also for trouble-shooting purposes and enables fault diagnosis to be carried out at the Royal Observatory Headquarters while simple tests are preformed by maintenance staff at the field site.

The software residing in the microprocessor controller is designed to:-

a) continuously monitor signals, if any, from the rain-gauge; and
b) when a rainfall signal is detected, update the cumulative rainfall total being displayed on the LED's and send a pre-defined data stream to the modem which frequency-modulates it and then outputs the modulated data stream to the telephone lines, via an isolation transformer.

Contact-bounce does not normally occur in rain-gauges equipped with 'mercoid' type of switches. For the rain-gauges presently used, however, the effect of contact-bounce has to be eliminated. This is achieved by including flip-flops between the rain-gauge and the microprocessor. Once activated by a rainfall signal, the flip-flops will remain inactive for a period of about 2 seconds. The time interval has been found to be long enough for contact-bounce to die down.

In order to protect the electronics against transients and lightning, a line filter and a large capacitor are provided at the AC power input and a 'tranzorb' surge suppressor is provided at the input from the rain-gauge. The site equipment is completely self-starting on the resumption of electricity supply after power failure.

Reception of rainfall information

The modulated signal at the receiving end of the telephone wires at the Royal Observatory is demodulated using a low-speed modem (Motorola MC14412). The restored serial data stream is fed into a microprocessor receiver controller (Mostek MK38P7Ø), the functions of which are to recognize the pre-defined bit pattern and then output a signal pulse to the microcomputer of the central control system.

Similar lightning and transient protection facilities as those at field sites are provided. The equipment is self-starting on the resumption of electricity supply after power failure.

Processing of data at the central control system

The central control system is built on the Mostek MD series microcomputer modules (Mostek Corporation 1980) and is designed around a Z-80 microprocessor (module CPU2A). A block diagram of the hardware congifuration is shown in Fig. 2.

The Z-80 system is timed by a clock module (Mostek BCLK) with a built-in battery back-up for at least five days, thus obviating the need to adjust the time after a power failure.

Using a Parallel Input/Output (PIO) module (Fig. 2) which provides 30 data lines (one for each rain-gauge), the microcomputer scans the output from each microprocessor receiver controller at regular intervals: once every 3/4 of a second. Another PIO module can be added to provide an extra 30 data lines with only slight modification of the software.

The software of the Z-80 system is written in Z-80 assembler language and resident on two 2-KB Erasable Programmable Read-only Memories (EPROM's) in the system. Dynamic memory has been allocated for each rain-gauge to keep account of rain fallen since the start of the following periods:-

(i) latest clock 5-minute period (i.e. since the 0th, 5th, 10th,....., 50th or 55th minute of the hour);
(ii) latest quarter period (i.e. since the 0th, 15th, 30th or 45th minute of the hour);
(iii) latest clock hour period (i.e. since the 0th minute of the hour); and
(iv) midnight last night.

Once the rainfall signal from a rain-gauge is detected at the central control system, a value of 0.5 (mm) is added to the rainfall total of the particular rain-gauge. At the same time, the time at which such information comes in will be written onto a computer-compatible tape (Fig. 2). The time corresponds to that at which the tipping of the bucket occurs, within an error of 3/4 of a second. Information stored in this manner enables the rainfall intensity during intense rainfall situation to be estimated in subsequent analysis (Donnelly 1979).

At the end of each clock 5-minute period, the rainfall detected by each rain-gauge (which is zero if there has been no tipping of the bucket) during the period will be transmitted to the main computer at the Royal Observatory, via a standard serial communication interface. The main computer processes the 5-minute rainfall data and generates various display products on a graphic terminal at the Central Forecasting Office for use by forecasters when considering warnings of floods and landslips.

At the end of each quarter of the hour, the 15-minute totals are monitored and an alarm will be sounded if the total at any one of the rain-gauges exceeds a specified threshold value.

Also, at the end of each quarter, the following are output to a line-printer :-

(i) the quarterly total;
(ii) the cumulative total since the start of the latest clock hour; and
(iii) the cumulative total since midnight.

The hard copy so produced is intended for back-up purposes. The transmission to the line-printer is done via a standard serial communication interface.

At midnight each day, all memory locations storing the rainfall totals are cleared.

The time of the central control system can be set using a video terminal connected to a duplex Universal Asynchronous Receiver Transmitter (UART) module (Fig. 2) through which interaction between the operator and the Z-8Ø system can be established.

Plate 2 shows the equipment set-up of the central control system.

Power supply to the central control system

Two Sola type 5Ø6ØØ uninterruptible power supply provides continuous 6ØØ VA AC power for at least 12 minutes. The central control system is self-starting on resumption of electricity supply after a complete power failure.

DISCUSSIONS

The availability of relatively inexpensive microprocessors simplifies considerably the electronic design of hardware. Electronic hardware conventionally built to accomplish some of the design functions is replaceable by microprocessor software. During the testing and implementation stages of the RDAS, efforts have been saved by simply making changes in the software instead of having to re-design the electronic circuitry.

In places where dedicated telephone lines are cost-prohibitive, it is possible to make modifications, mainly software, in the microprocessor transmitter and receiver controllers in order to accomodate the use of an ordinary dialling circuit as the communication link.

The essential parts of the central control system of the RDAS were implemented in November 1983 and eight rain-gauges became operational in December 1983. At the time of preparing this paper, the system has been running continuously for seven months. Apart from some minor modifications in the software during initial operation, there have been no major faults of any sort. The network will be expanded to 20 rain-gauges by 1985. The present system can accomodate a maximum of 60 gauges.

Work is in progress to incorporate into the microprocessor transmitter controller at the field site an automatic rainfall recorder (Lee and Chiu 1984) which stores rainfall data on solid-state Erasable Programmable Read-only Memory (EPROM). The power supply to the equipment at field site will be backed up by batteries. These measures will ensure that rainfall data will not be lost in case of telephone line failures and power stoppages at the field site.

ACKNOWLEDGEMENTS

I would like to thank Dr. H.K. Lam who gave useful advice on the overall design of the hardware.

REFERENCES

Donnelly, D.P.	1979	Digital Raingauge Recorder, J. of Appl. Met., Vol. 16, pp 205-207.
Lee, B.Y.	1984	A Real-time Rainfall Data Acquisition System, Royal Observatory Technical Note No. 72 (in press).
Lee, B.Y. and M.C. Chiu	1984	A Microprocessor-based Rainfall Recorder, Royal Observatory Technical Note No. 71.
Mostek Corporation	1980	MD Series Microcomputer Modules - Operations Manuals.

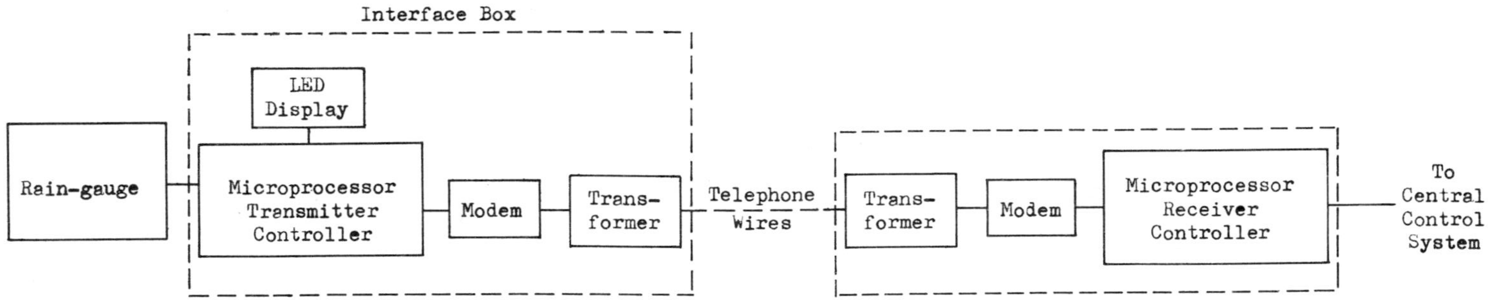

Fig. 1 Transmissicn and Reception in the Rainfall Data Acquisition System.

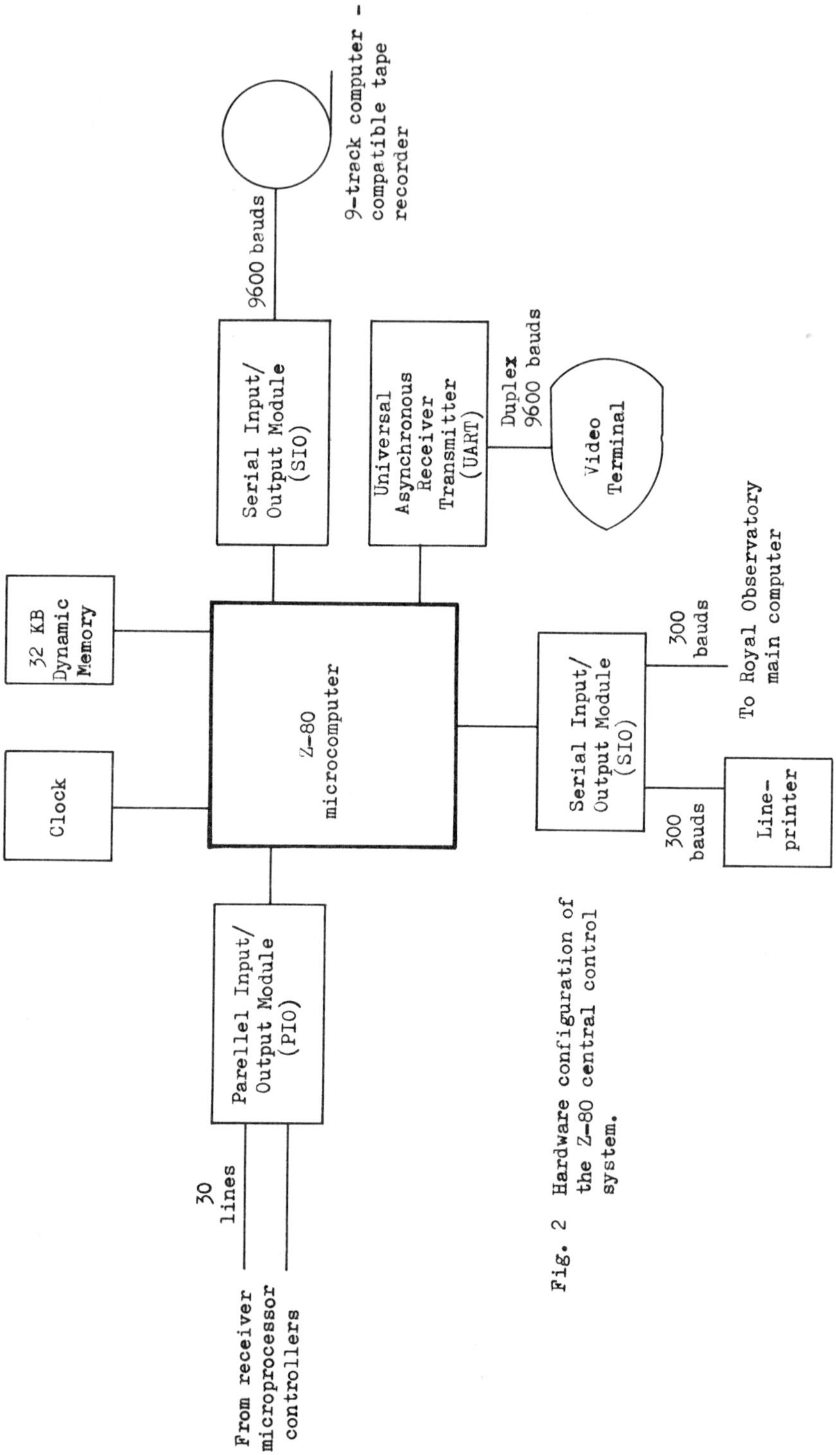

Fig. 2 Hardware configuration of the Z-80 central control system.

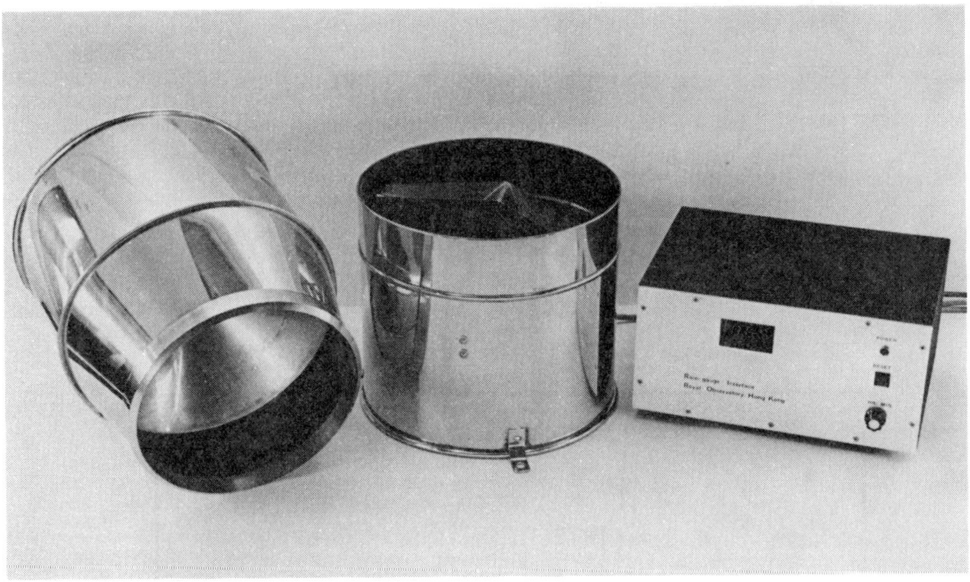

Plate 1. A tipping-bucket rain-gauge connected to an interface box.

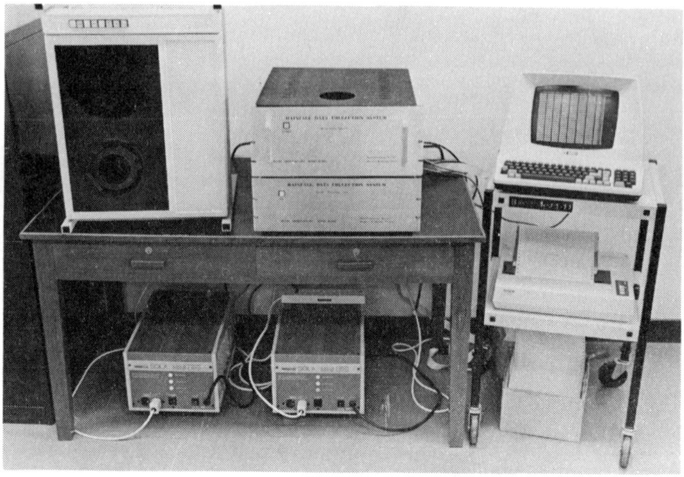

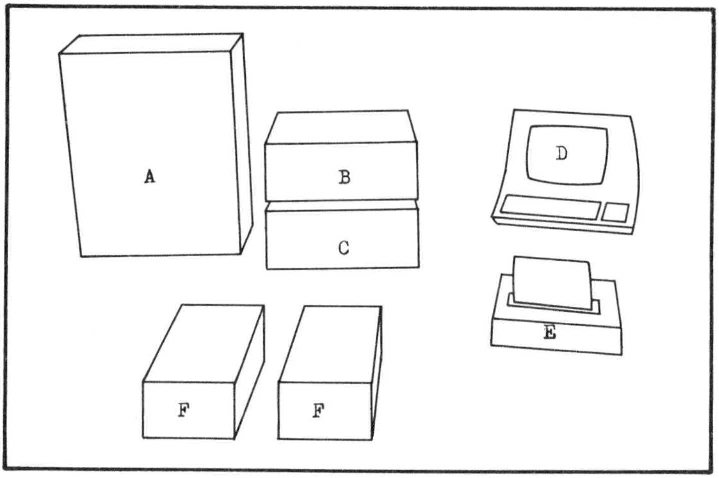

A. Computer - compatible tape drive
B. Microprocessor receiver controllers
C. Central Processing Unit (Z80 system)
D. Video terminal
E. Line-printer
F. Uninterruptible power supplies

Plate 2. The central system and peripherals.

DEVELOPMENT AND SUPPORT OF MICROPROCESSOR TECHNOLOGY FOR HYDROLOGIC DATA COLLECTION BY THE U.S. GEOLOGICAL SURVEY

Richard W. Paulson*

1.0 BACKGROUND

The emphasis on safeguarding the environment under the stress of agricultural, municipal, and industrial growth is placing agencies dealing with hydrology under increasing pressure to collect new and varied types of current data. There is promise that microprocessors, with their inherent flexibility, low power, and potential reliability, can assist in such data collection programs. In recent years, hydrologic data collection agencies have begun to use microprocessor technology for the automated collection and telemetry of hydrologic data and a variety of commercially available data loggers and data telemetry systems have been procured and used operationally in many countries.

This paper describes some of the technical and operational benefits and risks of the use of microprocessor technology in hydrologic data collection, and describes the U.S. Geological Survey's program to develop an advanced microprocessor-based hydrologic data collection system.

Historically, hydrologic data collection agencies have had relatively little experience with development of complex electronic data collection systems. In the developed countries the transition from mechanical data collection systems to electromechanical systems occurred in the last two decades. In some of the developing countries that transition is occurring now, while in other developing countries the transition from manual data collection to mechanical data collection is being made now. The transition to complex electronic data collection systems is being contemplated, or even initiated, by many in both developed and developing countries.

Many hydrologists and technicians in hydrologic data collection agencies in the developed countries have familiarity with microelectronic technologies, and commercial products using such technologies are being tested or used operationally with increasing frequency. Hydrologic agencies, whose management generally is staffed with hydrologists who are not experienced in planning applications with microelectronics are facing the need to invest large sums of money for planning and implementing new networks of advanced instrumentation because current automated instrumentation is outmoded and unresponsive to new data collection requirements.

The U.S. Geological Survey is using commercially available microprocessor technology systems to support hydrologic data collection and telemetry via geostationary satellites (see Paulson and Shope, 1984). The use of microprocessor technology has grown steadily over several years but still has not displaced the established electromechanical water level recorders that are the backbone of the Survey's automated data collection system.

The Survey has undertaken a program to upgrade the automated hydrologic data collection system it operates and replace the electromechanical recorders, water quality monitors and other hydrometeorological data collection equipment with a unified microprocessor-based system. There

*Instrumentation Coordinator, U.S.Geological Survey,Water Resources Division, Reston, Virginia.

World Meteorological Organization (ed.), Microprocessors in Operational Hydrology, 69-79.

are many technical and policy issues that are being addressed in this program to insure the successful development and operation of such an advanced hydrological data collection system.

2.0 IMPACT OF MICROPROCESSOR TECHNOLOGY ON HYDROLOGIC DATA COLLECTION AGENCIES

2.1 BENEFITS

There seem to be substantial potential benefits to the use of microprocessor technology for hydrologic data collection that can be derived from the flexibility and reliability it can provide. We sense this almost on a daily basis, as we are immersed in information about the electronics revolution. We have seen television broadcasts from the Moon, development of miniature radio and television sets, and the proliferation of home computers. We hear of the reliability, and reduced cost, power and size of electronic and microelectronic components. We also are aware of commercial products being produced for hydrologic data collection. From product specification sheets we see that data loggers are designed to collect data from many sensors, under numerous criteria, and that the data automatically can be evaluated, manipulated, statistically summarized at the hydrologic station and telemetered reliably via satellite or meteor trails from even the most remote locations.

Many data loggers seem capable of initiating alert messages or changing their data collection activities based on the hydrologic conditions they are monitoring. For example, some commercial Geostationary Operational Environmental Satellites (GOES) Data Collection Platforms (DCP) can initiate alert messages when water levels or rates of precipitation exceed specified levels. Under this mode of operation, routine data collection messages continue to be transmitted every several hours, while alert messages, transmitted at a different radio frequency, are transmitted every 15 minutes. Other DCP's are being used to test another mode of operation, which is known as random adaptive, that alters the interval between transmissions of hydrologic data in response to the status of hydrologic conditions. Both modes of operation allow communications systems to convey critical hydrologic data in a timely way without clogging such systems with nearly useless data when hydrologic conditions are static.

Microprocessor-based data systems also can collect and store data on a non-fixed time interval, somewhat analogous to the transmission schemes discussed above. A wide range of sampling schemes can be implemented by such systems to collect data at a high frequency during dynamic events and at a low frequency during more static periods.

These systems also can use data from several sensors to govern the collection and transmission of data, or command an alarm, or some other device. For example, data from water level and precipitation sensors can be evaluated automatically at a remote hydrologic station to initiate alert transmissions, or sound an alarm for flooding conditions. Commands to an automated sampler to collect water samples of urban runoff events also could be initiated by such a system that could monitor phenomena that are of too transitory for effective manual sampling.

Microprocessor-based systems seem to have the potential to operate under very harsh environmental conditions, and require low power. All of the attributes of microprocessor technology such as flexibility, reliability, low power, high data storage and telemetry can provide real cost and performance benefits to hydrologic data collection agencies.

2.2 RISKS

There are substantial risks to the use of microprocessor technology by hydrologic data collection agencies. One of the risks results from the great flexibiity of such technology. The objective of a hydrologic station is to extract hydrologic information from the data collected by the station. In the past, the information was extracted from data that were collected manually, then in analog form by a stripchart recorder and in recent years at a fixed interval of time by a mechanical recorder. There was relatively good institutional control over any automated data collection process because the technology was inflexible. Now a hydrologic station can be programmed to monitor sensors continuously and record data as a function of time, hydrologic conditions or a combination of both. Conceptually, each of a large number of hydrologic data collection stations equipped with an identical microprocessor-based system can be programmed to operate uniquely. They easily can be programmed to collect too much data and place an unwarranted load on data telemetry or processing systems, or the wrong data and miss the opportunity to collect the needed hydrologic information. Organizational control of data collection must be considered carefully in the use of such technology.

Another risk in the use of microprocessor systems is the loss of data due to data storage volatility. Many established automated hydrologic systems collect data on analog strip charts or punched paper tape, which are records that are durable if hydrologic agencies have developed proper standards for handling and archival. Microprocessor data collection systems record data on magnetic media, either in so-called solid-state storage or on magnetic tape, which are extremely volatile and reusable media. Many data systems can be programmed to automatically overwrite and destroy data that temporarily have been stored in the hydrologic station. Data can be destroyed or a microprocessor's operation disrupted by electromagnetic interference from generators, lightning, radio transmissions, etc. The vulnerability of data to improper retrieval and transport, or to transitory electromagnetic interference, can be a serious problem.

There are risks of commercially procured systems having poor reliability. Reliability can be expressed as a statistical mean-time-between-failure, as experienced by individual elements in a production lot of systems. It is difficult to develop and measure reliability standards and apply them to commercially procured systems, especially when procured in small quantities. System reliability results from the quality of design, components, manufacture and testing, and some commercial firms may minimize the cost of such quality assurance activities in order to offer their system at a low capital cost. Often the operating cost to the user of low reliability systems, i.e. numerous trips to hydrologic stations to retrieve failed systems, loss of data, personnel frustrations, etc., is far more than the apparent savings realized by the procurement of low cost systems.

There are substantial cost risks in maintenance, repair, and modification of microprocessor technology. Few hydrologic agencies have staff with a sufficient number and variety of electronic engineers, systems programmers, electronics technicians, nor laboratory facilities with programming and test equipment that are needed to support microprocessor technology. Even though hydrologic agencies may have proper human and technical resources, commercial firms are reluctant to share proprietary information about their products. This further lessens the opportunities for hydrologic data collection agencies to be self sufficient. Commercial firms that make elecronic components or hydrologic data collection systems normally have no legally binding obligation to continue to repair or provide these components and systems over a substantial number of years, and often drop these components and systems from their product lines as new markets and economic pressures develop. The demands for hydrologic data collection components and systems are small in comparison to other commercial demands for components and monitoring systems, and the pressures on commercial firms to innovate new components and systems and drop old components and systems is great. Often the hydrologic data collection systems that agencies buy become difficult or uneconomical to repair or maintain, and costly to replace.

A further risk to the successful operational data collection systems is the adoption of a technology for which staff is insufficiently trained. Often the demand by government agencies for skilled staff is hard to meet because government hiring restrictions and salaries place the government at a disadvantage with commercial firms with similar demands. It is common for new technologies to be adopted without a full understanding of the needed skills to keep them operating.

2.3 NEED FOR PLANNING

The brief discussion of risks above is not intended to stifle the use of another or new technology, such as microprocessor technology, but to alert hydrologic agencies to a few of many issues that must be examined as transition to new technology is made. The key to dealing with and minimizing risks is to properly plan the transition. The extent of the plan should be consistent with the economic investment that will be made in procuring and supporting a new system. If there is a need to buy and operate a very small number of units, which can be written off at the end of a project, then the planning effort can be modest and amount to a careful review of existing commercial systems. If there is a need for hundreds or even thousands of units of a new system, then the capital investment in the units, plus the operation and maintenance cost of their support, will be quite substantial and a thorough plan is warranted.

There is great benefit in conducting substantial planning studies of the need for, and requirements of, microprocessor-based hydrologic data collection systems. Such planning studies are intended to document to the greatest extent reasonable the expected benefits and risks in making the transition. Such documents can be excellent and necessary mechanisms for communicating within an organization the details of the transition, and maximize the possibility of creative ideas and valid requirements being communicated to program managers. Such communication can stimulate the initiation of associated policy and technical discussions elsewhere in the organization that can lead to the successful transition to a new system. Several potential activities that could be undertaken are discussed below.

Policy guidelines should be considered long before the implementation of a new microprocessor-based system to examine the legal and professional implications of adopting a new data storage medium, such as magnetic tape or solid-state storage. The use of such media can be highly successful if proper organization procedures are used to safeguard their use, procedures quite different from those in effect with an older technology. Such procedures should assure that nonvolatile archival copies are made of raw field data and audit trails to any human or machine changes to the data record are established. Modern computers can be used to manipulate data and a quality assurance program is needed to guard against their improper use.

Technical studies must be made of new data collection algorithms that are targeted for use in microprocessor data collection systems. In a large country such as the United States, with a large range of hydrological conditions that may be monitored by one agency, care must be taken to assure that an automated data collection scheme appropriate to one type of hydrologic regime also is appropriate if used in a second, and quite different, regime. Data collection schemes that are based on changes in water level, rates of change of water level, and time, as well as combinations of the three may not be applicable at all hydrologic atations.

Close coordination must be established between the group developing the field data acquisition system and the group responsible for the development and maintenance of office hydrologic data bases and processing systems. Both groups must keep in mind the end use of the data and the information products that are needed so that data records produced in the field and hand carried or telemetered to the office environment can be accepted and processed harmoniously by office-based computerized systems.

Strategy must be developed in specifying, and eventually repairing and maintaining the microprocessor-based field data acquisition system. If large quantities of units are needed then there is merit in procuring technical literature on the system from the system development firm so that the operating agency can be independent in its support of the system throughout its lifetime. A strategy must be developed to develop personnel and laboratory resources, or a system support contractor, that will enable the operating agency to meet its responsibility for system support. Strategy must be developed to assure that support and field personnel can feel comfortable with the new system. This can be done by proper training of personnel, designing the system so that it is straight forward to operate, or a combination of the two.

There are many other policy and technical issues that must be addressed in planning a new system, and this paper cannot exhaustively document them. The most productive way to identify them is through a series of meetings with all groups within an organization that potentially can be affected by the transition, by meetings with other organizations that have made similar transitions, and with commercial firms that potentially could provide development and manufacturing services. Such meetings should be well documented so that others who do not participate in the meetings can review the issues that were discussed and suggest others that need attention.

3.0 U.S. GEOLOGICAL SURVEY'S ADAPTABLE HYDROLOGIC DATA ACQUISITION SYSTEM PROGRAM

The U.S. Geological Survey is developing an advanced hydrologic data collection system to meet its mission as the principal water data collection agency in the United States. Working in cooperation with over 800 local, state and other Federal agencies the Survey collects data on the distribution and quality of surface and ground waters at over 60,000 locations in the United States. A principal component of this data collection program is the network of approximately 8,000 automated stream gaging stations where data collection instrumentation automatically collects data on water levels and water quality. Several years ago the Survey initiated a 7-year program to upgrade the generally electromechanical and outmoded data collection instrumentation that presently is in operation, to a new system, known as the Adaptable Hydrologic Data Acquisition System (AHDAS), that will employ microprocessor technology.

This system is intended to replace the 13,000 electromechanical punched paper-tape recorders, hundreds of water quality monitors, and hundreds of satellite, telephone and other telemetry components that are operating in the gaging stations. A conservative estimate of the replacement value of this equipment and miscellaneous commercially procured peripheral systems is approximately $20-25 million.

Because of the size of the network, potential capital investment and need to maintain the system for the maximum number of years possible at the minimum cost, the Survey has conducted extensive studies to plan, specify, and operate the system. These studies, and the attributes of the AHDAS system, are extensively discussed in Paulson et. al. (1984). The schedule of the AHDAS program, and the system's technical attributes, are very briefly described here.

3.1 AHDAS PROGRAM SCHEDULE

In the late 1970's the Survey conducted internal discussions about upgrading the entire water data collection system, which is comprised of field data acquisition instrumentation and a computerized water data storage and retrieval system. In 1980, the first formal study to upgrade the field instrumentation was initiated while, concurrently, studies were conducted to upgrade the office computerized system, which is known as the Distributed Information System and is described in Paulson and Shope (1983).

The Survey contracted with the Mitre Corporation in 1980 for a three-phased study intended to provide a basis for an improved field data acquisition system. The first phase was a systematic study of data system needs, the second phase was a review of available and emerging technologies, and the third consisted of a system research and development plan.

In the Fall of 1981, the Survey, with the continued assistance of Mitre, initiated an 18-month AHDAS system design study. The objective of this study was to develop a consensus within the Survey of the needed attributes of AHDAS and turn the consensus into a functional specification document that could be used for competitively selecting a system development contractor. The system design study also developed a host of support documents that stated the standards that the contractor had to meet in parts and materials, manufacturing, power, documentation, etc. Studies also were made of the environmental and electromagnetic conditions at Survey stations so that proper environmental standards and electromagnetic interference standards also would be conveyed to the development contractor. During the study, the Survey and the U.S. National Weather Service concluded an agreement to cooperate on the development of AHDAS as a hydrologic and meteorologic data collection system. Intensive discussions were conducted on tradeoffs between system cost versus reliability and versatility, and on the attributes of quality assurance programs that would be needed by the Survey and system development and manufacturing contractors. Finally, specifications were developed on system documentation so that the Survey and Weather Service would receive sufficient documentation to repair, modify or manufacture (by competitive contract) components of the AHDAS system. The system design study was concluded in the Spring of 1983, and a formal Request For Proposal to develop AHDAS was issued to industry in July 1983. The listing of specifications and background documents that were included in the RFP is found in table 1.

In January 1984, competitive negotiations were completed and a development contract was issued to the Teledyne Geotech Corporation of Dallas, Texas. As development contractor, Teledyne Geotech is charged with translating the Survey's requirements into physical prototypes, delivering 20 units to the Survey and Weather Service in the Spring of 1985.

These units will be field and laboratory tested to determine whether the system's functional specifications meet data collection needs, whether the units need the functional specifications, and whether environmental and reliability goals for the system have been met. Any failure to meet these tests will result in review and modification of the system and its supporting documentation.

I. Specifications

Number	Name
HIF-M-1	Minimum Configuration Field Component (MCFC)
HIF-I-3	Input Power Supply for the Field Component
HIF-P-3	Portable Field Interrogator (PFI)
HIF-C-6	Control Interface
HIF-F-1	Executive Software for the Field Component
HIF-P-4	Software for the Portable Field Interrogator
HIF-H-2	HYDRO Language Manual
HIF-A-6	Water Level Interface Module
HIF-A-7	Tipping-Bucket Raingage Interface Module
HIF-A-8	Analog Input Interface Module
HIF-A-9	Telephone Communicator Interface Module
HIF-A-10	Memory Module
HIF-I-2	Interface Module Software Guide
HIF-Q-1	Quality Assurance Program
HIF-E-1	Environmental (Rev. A)
HIF-A-3	Electromagnetic Compatibility
HIF-P-2	Parts and Materials Standards
HIF-D-3	Design and Manufacturing Standards

II. Background Documents

As Assessment of the USGS Field Station Environment
AHDAS Concept Report

Volume I, U.S. Geological Survey
Volume II, National Weather Service

Electromagnetic Compatibility and Electronic Field Monitoring Equipment Data Encoding Methods

AHDAS Functional Requirements

Volume I, U.S. Geological Survey
Volume II, National Weather Service

Table 1: Specifications and background documents that describe the AHDAS concept.

In the late 1985 or 1986 timeframe, the Survey and Weather Service will be in a position to issue a contract to begin to mass produce AHDAS units, which will be ready for field deployment in late 1986 or 1987. The complete modernization of the Survey's network of data collection instrumentation will proceed as rapidly as funding can permit.

3.2 AHDAS TECHNICAL ATTRIBUTES

This paper does not provide opportunity to fully describe the technical attributes of the AHDAS system listed in table 1, and the reader is directed to Paulson, et al (1984), or Walsh (1983) for a fuller discussion. As shown in figure 1, the AHDAS system concept consists of two principal units known as the Field Component (FC) and Portable Field Interrogator (PFI).

The FC is the element that resides in the hydrologic station and accesses, via a series of interface modules, a wide variety of measurement systems. Interfaces also are used to communicate data, via telephone and satellite radio. The heart of the system is an 8-bit microprocessor that, through a bus structure, can access these interfaces, memory, a communications port to the PFI and a test interface module. It will be powered by an external battery that meets the Survey's 12-volt battery standard. It will lack all control switches, save one that can be used to trigger a display of current data, which will allow an observer to make a simple check of its performance. The FC will operate under control of a user-defined algorithm, written in a special high-level language known as HYDRO, that will allow each channel of data to be monitored and data stored according to the criteria specified for that type of data.

Communications with the FC normally will be done via the PFI, which is a transportable unit the field technician will carry to the hydrologic station where it will be electrically connected to the FC. Communication between the two units will allow data to be retrieved by the PFI and permit the FC to be programmed and tested by the PFI. After the technician is satisfied that the FC data have been retrieved, and the FC is properly configured, the PFI isdisconnected and used to convey data back to the office computer system, or to a convenient telephone to download the data to the office computer system. The PFI also will employ a microprocessor and be powered from an internal battery. The PFI will have a full keyboard and display capability for digital data, as well as an analog display of data.

3.3 AHDAS SUMMARY

It is common for the development of any complex system to take 10 years. AHDAS, from the time that it was identified that the existing data collection system needed upgrade, until it is scheduled to begin field deployment, will take approximately seven years. Throughout this period, the Survey has, and will be, conducting associated studies and reviews of ancillary issues. These include the policy and legal implications of making the transition to a magnetic recording medium from a punched paper-tape recording medium. Studies are being made of moving from a fixed time interval of data recording to event-driven sampling schemes. Proper interfacing of AHDAS with the DIS will be addressed so that the two major elements of the data system can harmoniously be integrated. Intensive studies are being conducted to assure that the Hydrologic Instrumentation Facility (HIF), which is the Survey's principal hydro-

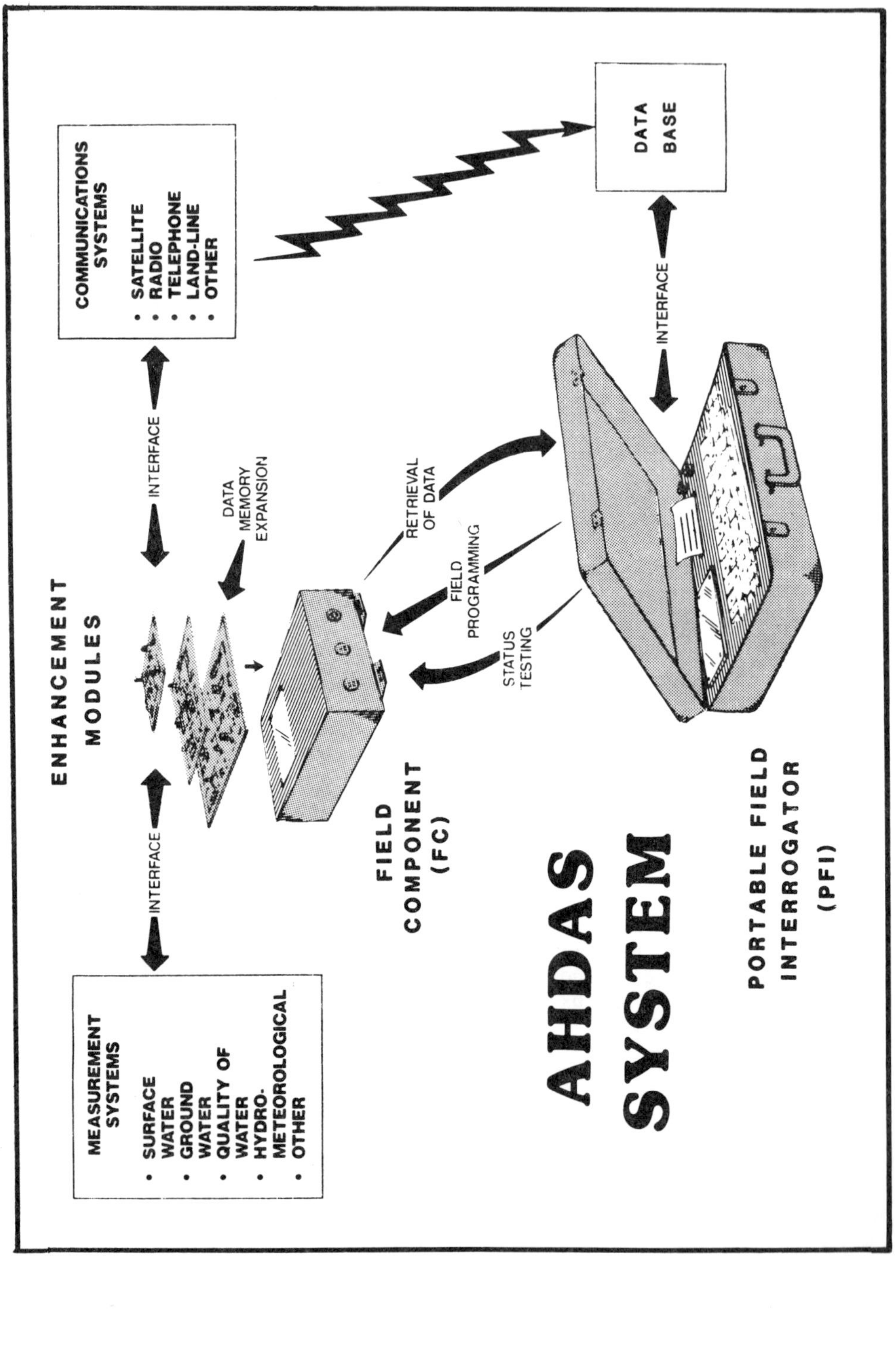
MEASUREMENT SYSTEMS
• SURFACE WATER
• GROUND WATER
• QUALITY OF WATER
• HYDRO-METEOROLOGICAL
• OTHER
INTERFACE
ENHANCEMENT MODULES
INTERFACE
COMMUNICATIONS SYSTEMS
• SATELLITE
• RADIO
• TELEPHONE
• LAND-LINE
• OTHER
DATA MEMORY EXPANSION
FIELD COMPONENT (FC)
RETRIEVAL OF DATA
FIELD PROGRAMMING
STATUS TESTING
DATA BASE
INTERFACE
AHDAS SYSTEM
PORTABLE FIELD INTERROGATOR (PFI)

logical instrumentational support facility, is properly equipped and staffed to support the maintenance, repair, and procurement of AHDAS units, subassemblies and components. (Wagner, 1984)

4.0 CONCLUSION

The risks of the transition to new microprocessor technology for hydrologic data collection instrumentation can be minimized through thorough planning. The myriad of technical, policy, legal, and cost issues that affect such a transition must be thoroughly explored and discussed with individuals within and without of the agency that is planning the transition. Such discussions assist the development of a consensus on the performance, cost, and reliability of a new system and on the strategy to support it. Only through the development of such strategies can the risks of such a transition be minimized and the benefits maximized.

REFERENCES

Paulson, R. W., Billings, R. H., and Cherdak, A. S., 1984, The Adaptable Hydrological Data Acquisition System, in New Advances in Hydrology, Adam Hilger Ltd., Bristol, U.K.

Paulson, R. W., Shope, W. G., Jr., 1983, Development of a Distributive System for Handling Real-Time Hydrologic Data Collected by the U.S. Geological Survey, IAHS International Symposium on Remote Sensing and Remote Data Collection, Hamburg, FRG.

Walsh, Peter D., 1983, Development of Advanced Data Logging Instrumentation at U.S. Geological Survey - Adaptable Hydrologic Data Acquisition System, Northwest Water Authority, Warrington, U.K.

Wagner, C. R., 1984, Hydrologic Instrumentation Facility, in New Advances in Hydrology, Adam, Hilger Ltd., Bristol, U.K.

TELEMETERING HYDROMETEOROLOGICAL NETWORK IN MENDOZA, ARGENTINA ONE YEAR OF EXPERIMENTS AND RESEARCH

P.C. Fernandez*; Osvaldo Roby**; Luis Fornero**; Jorge Maza**

1. INTRODUCTION

In recent years, one advance in scientific hydrology has been the development of sophisticated mathematical models, the use of which has become common in the design of hydraulic works, as a result of the enormous progress in, and widespread use of computer technology.

Similarly, the development of increasingly precise automatic data-acquisition equipment at decreasing cost has had a decisive impact on the instrumentation in catchments and on the practical possibility of having operational real-time systems for use either as flood warning networks or for more efficient water-resource management [2, 4].

These facts are part of the reason why the Andean Regional Centre of INCYTH is currently setting up a telemetering network on a pilot area of 600 km^2 to the west of the city of Mendoza in the foothills of the Andean Cordillera.

Brief reference is made in this paper to the features of the aforementioned network, and to the results obtained after the first few months of the network's operation. The first stage of the network has now been completed.

2. BACKGROUND AND LOCATION

For several years, the Instituto Nacional de Ciencia y Técnica Hídricas (INCYTH) (National Institute for Water Science and Technology), through its Andean Regional Centre, has carried out hydrological research in the foothills of the Andean Cordillera, west of Mendoza city (Figure 1).

For the purpose of the studies, a network of eight conventional hydrometeorological stations was set up over approximately 600 km^2 covering nine small contiguous basins. After eight years of operation, analysis of a great many storms and the study of radar images, it was concluded that because of the rather small size of convective cells (they vary between 5 and 15 km in diameter), as well as their erratic distribution, a denser, automated network was required in order to have adequate temporal and spacial coverage of the convective storm fields and their relationship to floods in torrents.

The intensity and short duration of precipitation, together with the extremely short concentration time of the catchments (less than one hour) were limiting factors to be considered in the selection of the data collection system to be installed. From the analysis of the various existing surface-based telemetering systems, the one selected was that developed by the National Weather Service River Forecast Centre of California, USA, known as the event reporting system [2, 5, 7, 8, and 9]. It was selected for the following reasons:

---

* Research Director of the Andean Regional Centre of INCYTH. Mendoza, Argentina.

** Research workers in the Andean Regional Centre of INCYTH. Mendoza, Argentina.

World Meteorological Organization (ed.), Microprocessors in Operational Hydrology, 81-90.

- Low cost;
- Simplicity, ease of installation and maintenance;
- Instant response, without dead time, which makes it the only genuinely real-time system for torrential basins;
- Compact, self-contained design which makes it more vandal-proof;
- No need to synchronize the raingauges, since the whole system functions on the basis of a single clock in the central station computer.

3. NETWORK OBJECTIVES

The main purpose of the network is to provide detailed summer storm data which the Centre can apply in its hydrological studies and its studies on the calibration and verification of the parameters of mathematical rainfall-runoff models for use in the hydrology of torrential basins and urban areas.

Another major objective is to set up a hydrological warning system for river floods, in this instance for the protection and management of the river-flood control canals in the Greater Mendoza area.

A third, indirect, but no less important objective is that of installing and operating, for the first time in Argentina, a pilot network of this type that demonstrates the benefits of such technology for the automatic recording of hydrometeorological data and for real-time river runoff forecasts.

4. PLANNING OF THE SYSTEM

Figure 2 shows the location of the network's stations. The network is made up of 15 remote raingauge stations and one flow metering station.

On the basis of earlier studies [1, 6 and 8], in order to have a good definition of convective cells, there had to be 1.5 to 3-km intervals between the stations.

This proved an economic impossibility for the 600 km^2 area. Hence an area of 36 km^2 was selected, that included a pilot catchment of 5 km^2 where 12 raingauge stations were placed (one every 3 km^2) forming a three by four grid (Figures 2 and 3) and a flow metering structure was set up, calibrated according to a physical model using level telemetering through a bubble gauge [7, 8 and 11].

The rest of the area is being set up with an operational purpose, to function as a hydrological warning network for the protection and management of the drainage canal system of Greater Mendoza.

For this purpose, on the basis of experience acquired in the 82/83 and 83/84 seasons, it was decided to cover new areas and to increase the number of raingauge stations to a total of 27, and of stream gauging stations to 3, as indicated in Figure 3.[7].

Regarding economic studies, not only did the system chosen prove three to seven times less costly than the others considered [7, 8] but its low-cost installation and maintenance made the difference even larger.

5. OPERATION

The average efficiency of the system's operation over these last few months amounted to 99.5%. The stations are, in normal circumstances, examined once a year and all that is required is a change of battery. The rare breakdowns that have occurred were caused in two stations by acts of vandalism (antennae broken) with an eight days disruption in service for one station and a two-day disruption for the other.

In another two stations, the electronic unit failed, but this was quickly repaired on the spot replacing of the faulty card by a spare. The faulty card was then repaired in the workshop so the station was not out of service, for any longer than the time necessary to discover the failure and to reach the station (2 days in each case).

The batteries have lasted for one year without replacement in 14 out of 16 stations, which resulted in disruption of service for 12 days in one case and 20 in another because of difficulties of access. In all, 46 station-days were lost.

Thus far, the 16 stations have operated for 547 days, i.e. 8,752 station-days with a breakdown percentage of 0.5%.

The rainfall sensors, calibrated before the station went into service, have been compared with raingauges in the conventional network which use a strip-chart recording mechanism and which were kept in service for this purpose.

The response of the network during floods is instantaneous, because of the event reporting system. Thus immediate information is obtained about the location, intensity and track of the storm, without the problems that usually arise in high-intensity convective storms with systems with polling from the central station or systems that transmit information at regular intervals.

Signal interference caused by the simultaneous arrival of two or more signals from remote stations into the central station, even in intensive rain, is so low for a network of 16 stations as to be insignificant. This in principle confirms the figures mentioned in reference [5] of 8% interference in 100 stations with a uniform rainfall of 50 m/hour.

The operating cost of the remote stations in this initial period was very low and it has not been possible to estimate it because the travel and replacement of parts fell within the Centre's routine operations. However, the maintenance cost of the computer of the central station has amounted to US$ 200/year.

The network's data collection process is managed through a microcomputer with a 64 K byte memory, 2 diskette drives and a printer (Photo 1).

The software used is the ALERT program of the N.W.S. of the United States, which controls an RS232 postwhich receives the data from the network and stores them on diskette. Moreover, the program provides for consulting the information stored for each sensor or group of sensors.

At present, the program's stored data remote consultation facilities are not being used, although it is planned to link the data collection system to a bigger computer where rainfall-runoff models will be set up in order to obtain runoff forecasts.

Given that the version of the data collection programme does not provide for permanent storage, programs have been developed for storage and consultation of a data bank which is periodically updated.

Figure 4 gives various options whereby the collected information can be retrieved.

6. RESEARCH

Although the number of storms recorded during this short period is insufficient to permit generalization of the results, the information obtained has already shown how useful the system is for operational and research purposes.

In relation to the study of convective storm fields, the records show that the isolated nuclei have diameters between 5 and 15 km, and that their distribution over the area is completely erratic in the north-south direction, that is, in the direction of the contour lines.

With relation to their east-west distribution, it was found that the highest values occurred below 2,000 m.a.s.l. and there was no evidence that the same storm could show significant values above or below that level.

Figures 5, 6 and 7 give total values for the storms of 17 and 30 December 1983 and of 20 March 1984.

With relation to research on calibrating and verifying mathematical rainfall-runoff models, the information is being used to obtain infiltration and interception loss parameters and parameters for the unit hydrograph in the HEC-1 model of the United States Army Corps of Engineers [10].

As already mentioned, data collection and processing are done in real time. The information is instantly obtained by simply interrogating the microcomputer's file. Regardless of any operation going on in the computer, the information received takes absolute priority for input and filing on diskette.

The system works in "genuine real" time without dead time, and this is truly a key element in torrential basin warning systems where concentration times are measured in a few minutes (flash floods).

The event reporting system for rainfall and river level data makes it possible to transmit directly to the central station's computer at the instant the measurement is made. Every tip of the raingauge bucket (1 mm) or every stage variation at the gauging station automatically activates the transmitter that sends the signal in real time to the central station. The rainfall information is the accumulated value, so that if any signal is lost through interference, the next signal will be the aggregate of the last two. In other words, the total value of the rainfall is not lost, all that is lost is the information on the instant in time at which the rainfall millimetre precipitated.

The data from the other sensors in the network (temperature, humidity, solar radiation and so forth) are transmitted at regular intervals (varying from 3.6 minutes to 12 hours).

With current software, the information is recorded at minimum intervals of 3 minutes. In intense rain, e.g. during flooding, this is a very long time, as can be seen in Tables 1 and 2 where the accumulation of transmissions may be noted but are assigned to a single interval without any possibility of identifying their real distribution; this shortcoming is more sensitive for flows. Next season, with the incorporation of modified (enhanced alert) software the information will be recorded every second.

One basic requirement for using the flow forecasting system in real time is the instantaneous operation of a mathematical rainfall-runoff model; this will be met through the use of a bigger computer, which is to arrive shortly.

When the computer is received, new software will be set up with other options making it possible to handle information in a more useful and flexible manner not only for real-time operation, but also for the analysis and statistical processing which are basic requirements in hydrological research. Moreover, models will not be used only in real time but also in deferred time, e.g., for obtaining historical data by station or groups of stations in standard format (as they are now being received) or for pre-established dates or intervals. Historical data will be obtained in the form of maps for a pre-established date or interval, and it will be possible to key data in manually, an operation that is very useful for calibrating mathematical models.

7. CONCLUSIONS

From the point of view of hydrological research, the need is to supplement the network of 12 stations in the pilot area and install one or two stations to the north in order to obtain reports on conditions on the edges of the nuclei centred over that zone.

The present system of recording rainfall or level data at the central station takes a minimum of three minutes and must be reduced at least to one minute in order to have an undistorted determination of both rainfall and flows.

Another two flow measuring stations must be set up to analyze the response of the larger catchments which are totally or partially affected by storms.

In order to obtain a hydrological alert system, the following is required: to supplement the network, taking it into areas at present not covered (Fig. 3) both in the rural and suburban area to the east of the stations now in service and even into the urban area; to duplicate the data-collection, processing and archiving equipment of the central station; and to set up an automatic power generator in order to avoid disruption of service during power cuts.

The aim is to put into operation a calibrated rainfall-runoff model on line with the data-acquisition system in order to obtain discharge forecasts at pre-established critical points.

Telemetering networks are basic elements in current hydrological research and form the essential linkage between the mathematical rainfall-runoff models at present being developed and their application both to works design engineering and to real-time systems operation.

The development of rainfall and discharge forecasting technology using computerized radar or sectorized weather satellite images (eg., real-time) calls for networks of this kind for their calibration and verification.

REFERENCES

1. Amorocho, J. (1980). Report on recommendations for a hydrologic study program. INCYTH-Centro Regional Andino. Mendoza, Argentina.

2. Burnash, R. & Twedt, T. (1978). Event reporting instrumentation for real-time flash flood warning. Conference on flash floods - hydrometeorological aspects. Los Angeles, California, USA.

3. Burnash, R.J. (1979). Automated precipitation measurements. National Weather Service River Forecast Centre. Sacramento, California, USA.

4. Burnash, R.J. & Ferral, L. (1982). Examples of benefits and the technology involved in optimizing hydrosystem operation through real time forecasting. National Weather Service, California-Nevada River Forecast Center. Sacramento, California, USA.

5. Burnash, R.J. (1982). Real-time rainfall-runoff forecasting. Today's technological revolution in flood protection. Spring meeting of the American Geophysical Union, 1982.

6. Brun, S., Pedrani, A. & Fernandez, J. (1981). Medición de caudales y ajustes de modelos hidrológicos en Divisudero Largo (Flow metering and adjustment of hydrological models). CRICYT-IADIZA-INCYTH. Informe de Avance. Mendoza, Agentina.

7. Fernández, P., Amorocho J., Roby H. and Fernandez J.M. (1983). Hydrometeorological network for real-time data collection in southern hemisphere arid area. Proceedings of the Hamburg Symposium. August 1983. IAHS, publ. No.145.

8. Fernandez, P.; Amorocho, J.; Roby, H.; Maza, J.; Fornero, L., (1983). Planning, installation and operation of an event reporting data system in an arid zone of Argentina. Technical Conference on the Mitigation of Natural Hazards through Real-time Data Collection Systems and Hydrological Forecasting, WMO-NOAA. Sacramento, USA. 19-23 September 1983.

9. National Weather Service (1981). Western Region Hydrologic Service Division. Automated local evaluation in real time. Salt Lake City, Utah, USA.

10. US Army Corps of Engineers (1981). HEC., HEC-1. Flood hydrograph package users manual. Davis, California, USA.

11. Universidad Nacional de San Juan (1982). Instituto de Investigaciones Hidráulicas-Estudio sobre modelo físico vertedero aforador Divisadero Largo. San Juan, Argentina. (Hydraulics Research Institute - Study on a physical overflow meter model - Divisadero Largo).

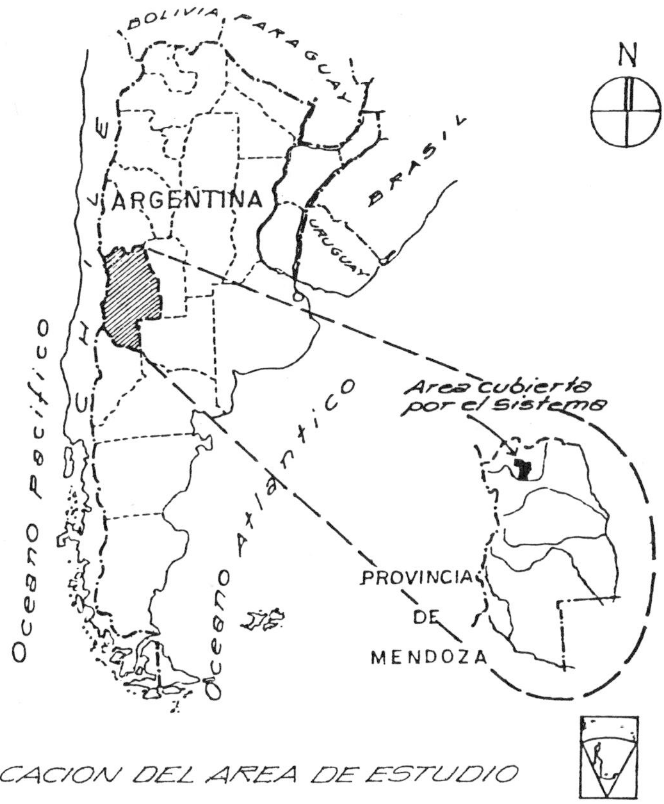

Figure 1. Location of study area

Photograph 1

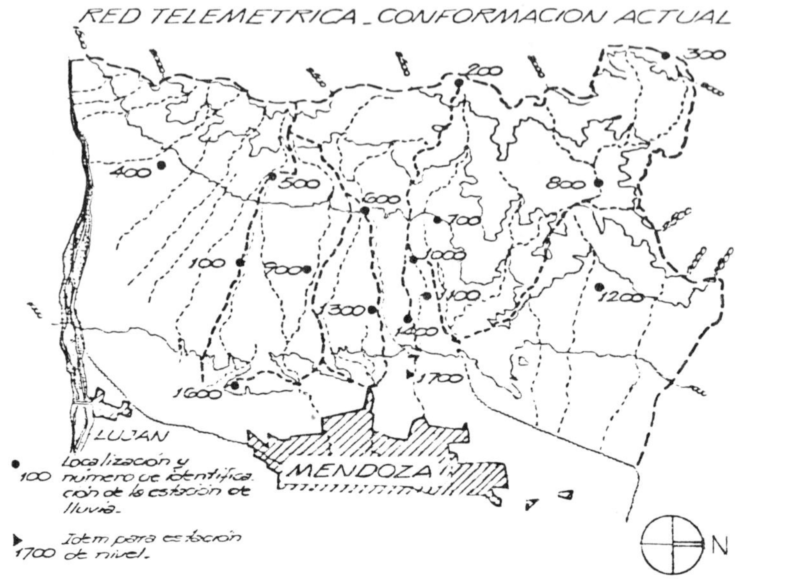

Figure 2. Telemetering network - present set up

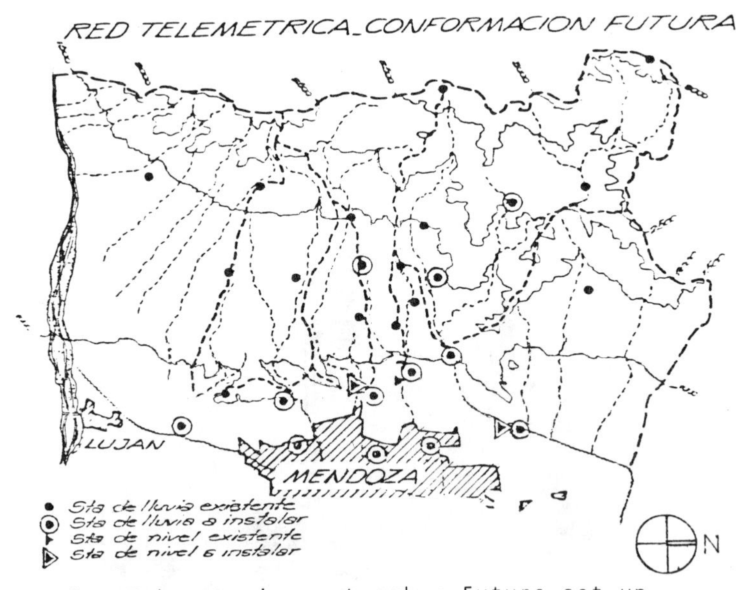

Figure 3. Telemetering network - Future set up

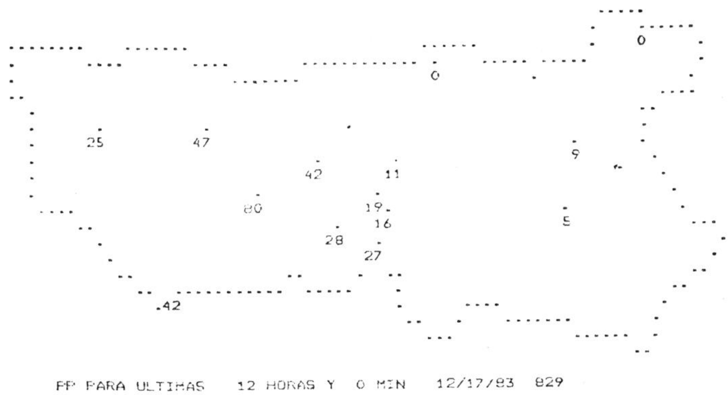

Figure 4. ALERT system displays

ESTACION = 900 DIA = 1216

1245	38.00	2351	40.00	2354	42.00	2357	44.00	2357	46.00
2351	39.00	2354	41.00	2354	43.00	2357	45.00		

ESTACION = 900 DIA = 1217

0	47.00	48	63.00	136	77.00	218	93.00	242	108.00
0	48.00	51	64.00	139	78.00	218	94.00	248	109.00
3	49.00	54	65.00	139	79.00	218	95.00	251	110.00
3	50.00	57	65.00	142	80.00	218	96.00	254	111.00
9	51.00	103	66.00	145	81.00	221	97.00	300	112.00
12	52.00	112	67.00	148	82.00	221	98.00	315	113.00
12	54.00	115	68.00	148	83.00	224	99.00	336	114.00
15	55.00	121	69.00	151	84.00	224	100.00	639	115.00
30	56.00	124	70.00	154	85.00	227	101.00	645	116.00
39	57.00	127	71.00	157	86.00	230	102.00	654	117.00
42	58.00	130	72.00	200	87.00	230	103.00	718	118.00
45	59.00	130	73.00	206	88.00	233	104.00		
45	60.00	133	74.00	212	90.00	233	105.00		
45	61.00	133	75.00	215	91.00	236	106.00		
45	62.00	136	76.00	215	92.00	239	107.00		

Table 1

ESTACION = 1700 DIA = 1216

1851	0.00

ESTACION = 1700 DIA = 1217

233	3.40	236	20.40	242	37.40	318	34.00	515	17.00
233	6.80	236	23.80	245	40.80	333	30.60	703	17.00
236	10.20	239	27.20	245	44.20	354	27.20	706	13.60
236	13.60	239	30.60	248	40.80	412	23.80	757	10.20
236	17.00	22	34.00	306	37.40	448	20.40	854	6.80

ESTACION = 1700 DIA = 1218

727	6.80	1909	3.40

Table 2

TORMENTA DEL 16/17 DE DICIEMBRE DE 1983 _ 23^{00} hs. a 03^{00} hs.

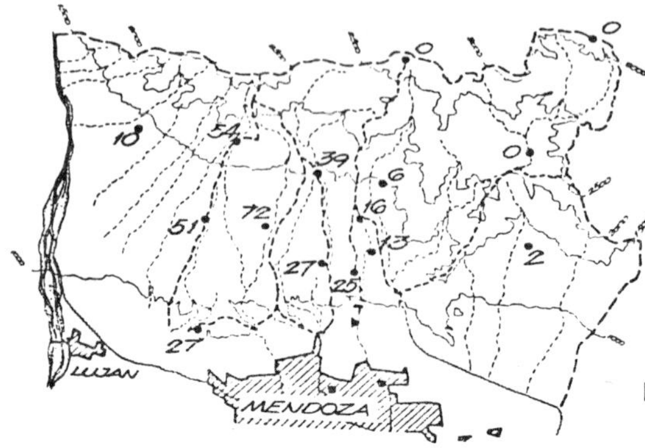

Figure 5

TORMENTA DEL 30 DE DICIEMBRE DE 1983 _ 23^{15} hs a 3^{15} hs

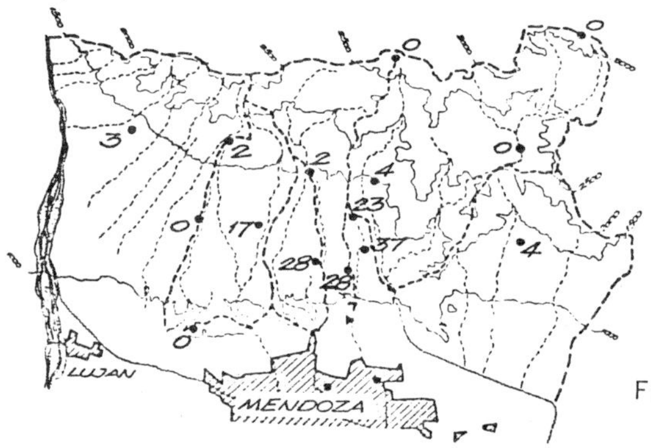

Figure 6

TORMENTA CONVECTIVA CON VARIOS NUCLEOS _ 19/20 - 03 - 1984

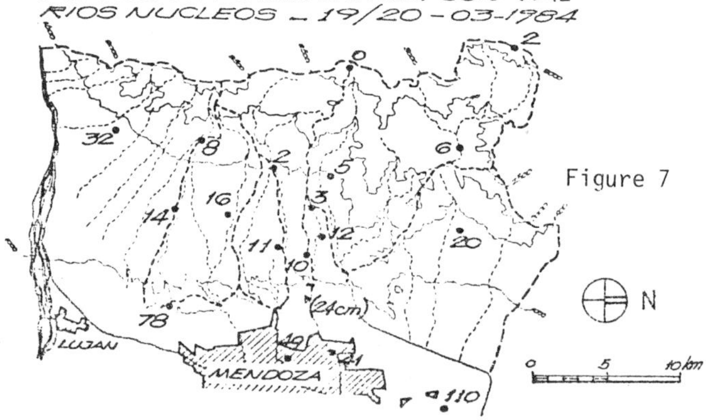

Figure 7

A LARGE SCALE ENVIRONMENTAL DATA COLLECTION - DISSEMINATION SYSTEM USING A MINICOMPUTER NETWORK

ION DUMITRASCU*

1. INTRODUCTION

The growing awareness of the ecological effects of industrialization and the environmental dependency of human activities has stimulated new monitoring activities. Environmental parameters such as: barometric pressure, air temperature, humidity, wind speed and direction, precipitation, solar radiation, snowfall, radioactivity, pollution, and water levels are measured by human operators or by automatic stations with associated sensors, transmission and processing of the data obtained, resulted in minicomputer oriented data collection systems. To collect environmental data from a large geographical area, many such acquisition systems must be connected in a hierarchical, flexible and highly reliable system structure. The purpose of the system is to collect, process and transmit environmental data in real-time to end-users to be used in meteorology, hydrology, flood forecasting and warning, pollution control, river management, agriculture and transport.

2. OUTLINE OF THE SYSTEM

Relating to the arrangement provided for the nation-wide network activity involving weather and river stations, the environmental data acquisition and dissemination system will be monitored automatically, by many automated subsystems. A supervisory minicomputer for each subsystem perform data collection by automatic polling of the individual stations and terminals. The raw data are

* Project Chief, Research Department, Central Institute for Management and Informatics, R 71316 Bucharest, Bd. Miciurin 8-10, ROMANIA

World Meteorological Organization (ed.), Microprocessors in Operational Hydrology, 91-96.

processed into a standardized meteorological message format (SYNOP, HYDRA, TEMP, CLIMAT, described in World Meteorological Organization recommendations) and stored.

Centralization of the collected information through the subsystem channels will be achieved at the head office, where the information will be processed to produce messages for national and international exchange. The system will provide data for international data exchange according to the rules of the World Meteorological Organization, described in the Manual on the Global Telecommunication System, Document WMO - No. 386.

End-user requirements for the systems are: flexibility - system arrangement and the application software should enable on-going network development without affecting the main functions; adaptability - system facilities can be modified as new requirements appear; reliability; more cost-effective development; development in short time.

3. SYSTEM STRUCTURE

The environmental data collection - dissemination system is a hierarchical structure consisting of many data collection subsystems and one data concentration system connected to a large scale data processing system. At the low level are the minicomputer oriented collection systems. These minicomputers have few peripherals, terminals for local data introduction and many communication controllers for automatic stations and telecommunication links. This is a dedicated message switching system with extensive capabilities for the processing of the meteorological data and consists of a minicomputer with associated peripherals (mass storage devices such as discs and tapes, terminals, visual displays, line printers) and communication controllers for synchronous and asynchronous operation. Initially the connection is in a star network. Depending on end-user requirements and reliability needs the system may later be connected in a distributed network.

4. SYSTEM CHARGES

At the data collection level the primary function is data collection from the automatic stations and terminals, and transmission to the concentration level. That means: automatic time-controlled polling of the individual stations; technical status information analysis for each automatic station to check correct operation; data conversion using sensor associated algorithms; data validation; data processing into a standardized meteorological message such as SYNOP, HYDRA, CLIMAT, TEMP, etc. A secondary function, at this level, might be end-user information through information bulletins and user warning in dangerous situations detected by data comparison with normal limits. The information bulletins and the warnings are organized as addressed messages with the format of the meteorological messages and introduced in the normal flow of the message for switching.

The concentration level is responsible for: data reception from the local supervisory minicomputers; data storage and retrieval; development of new messages and information bulletins; international data exchange using meteorological telecommunication procedures for the Global Telecommunication System; data exchange with the large scale computer dedicated to environmental data processing and storage (data base manager).

5. HARDWARE AND SOFTWARE FACILITIES

The Romanian built COMETA automatic stations and terminals are used for data acquisition. COMETA automatic stations are designed to operate unattended in harsh environmental conditions for periods of many months. Telecommunication between the automatic stations and the minicomputer is by wireless equipment. Common error protection methods are used for data transmission. Each radio line requires a line from the asynchronous multiplexer and suitable modems.

The terminals are normal teletypes used for the introduction of the data received from human operators. For the remotely located terminals, telecommunication links used are telephone common or private lines.

The minicomputers used are the Romanian I 100/I 102 hardware systems with main memory between 64 Kbytes and 256 Kbytes and peripherals consisting of mass storage devices (discs and tapes), terminals, line printers, displays and communication controllers for synchronous and asynchronous operation.

Communication between minicomputers and the international data exchange is over synchronous telecommunication links using private telephone lines.

Data dissemination to end-users is generally over

undisciplined asynchronous links using telex, common, or private lines and suitable adaptors to the asynchronous multiplexer of the minicomputer.

The operating system used is the AMS (Adaptable Multifunctional System) designed in Romania. AMS is a real-time multi-programming operating system which provides real-time, event-driven program execution in a multi-programming environment. The AMS operating system can support MININET network software, which can be used as a component of distributed networks and communications networks, in point-to-point, star or mesh network topologies.

MININET facilities are: intertask communication - allows two tasks in same system or in remotely located systems to create a data path for transfer of information between them; network resource access - device sharing, remote file transfer, remote file access, virtual terminals; remote task control - allows control of the execution of installed tasks between two nodes; down-line task loading - allows transfer, loading and execution of a program at a remote system; down-line system loading - allows transfer, load and start an AMS system at a remote adjacent node; interterminal communication - allows a user at a terminal to exchange messages with another terminal at any node in the network.

6. CONCEPTION OF THE APPLICATION SOFTWARE

The use of a distributed minicomputer network as a solution for the environmental collection - dissemination system development was selected concerning its utility in geographically dispersed applications requiring reliability and high computing power.

Using the minicomputer network software, which gives a general set of tools (data communication, file sharing, intertask communication), the application software has been designed in a generalized modular fashion, in order to accommodate future adaptation to new requirements.

The application system was designed to meet the major end-user requirements: configuration flexibility, functional adaptability, reliability, easy maintenance, and cost-effective development.

The following elementary functions were identified: data acquisition from the automatic stations, data collection from terminals, data storage and retrieval, data checking, new message development, communication management, data accounting, and operator communication. For these functions independent application software modules, called tasks were developed. All modules were designed so that modifications of any of them does not need modifications in other modules. Each task can have one or many inputs and outputs. Each task receives data from another task or from communication lines, processes them and transfers them to another task. This application software design was selected in order to simplify future adaptation to new requirements. A system generation process is used to select the tasks and the particular form of each to be used in a network minicomputer. Another advantage of this solution is fault isolation at task level and easy debugging.

The data must be processed successively by many tasks. To reduce the number of accesses to the disc, where the data are stored, the selected solution was the transfer of the data directly between the source task and destination task using the intertask communication facility of the minicomputer network software. The synchronization of processing tasks is assured by the network software, because a transmission to another task is possible only when the destination task is ready to receive data.

Another facility of the network software is the on-going replacement of some application tasks by new versions. These tasks are so designed that they can connect and disconnect all their connections with other tasks in proper conditions. Using the operating system facilities the disconnected task can be replaced dynamically by another version and the new active task restablish its connections.

The communication of data between two remote minicomputers is implemented as a communication between the tasks for data storage and retrieval (the main task).

All the communication problems are solved at the level of the network software (not by the programs of the application software). The use of the network software facilities is also useful for easy maintenance at the level of the environmental data collection - dissemination network. The new software modules such as new versions, newly generated modules, etc., are developed in one minicomputer from the network, by the application software team, and are transmitted to other minicomputers using the down-line task loading facility of the network software.

The real-time character of this application requires the application software to be memory resident, along with the network software of and the operating system. Because of memory restrictions it may happen that not all the tasks necessary for one subsystem can reside in a single minicomputer. In this case the application software can be divided between two locally interconnected minicomputers because the same intertask communication is used for one minicomputer or for two different minicomputers. This is another advantage of the minicomputer network usage.

7. REMARKS

The application software runs on CORAL or INDEPENDENT Romanian minicomputers, but also on any of the PDP-11 minicomputers (Digital Equipment Corporation) using RSX - 11M V3.2 operating system and DECNET - 11m minicomputer network software, with small modifications required by any device drivers.

The task for data acquisition from the automatic stations can be modified to be used for other type of stations.

Finally, this concept can be used to develop geographically dispersed environmental data collection systems based on other minicomputers with the network communication facilities.

8. ACKNOWLEDGEMENT

The Institute for Technical Engineering for Automation (for automatic stations and communication equipments), the Research Institute for Computer Technique (for hardware devices and operating system), the Institute for Meteorology and Hydrology (as end-user) and the Central Institute for Management and Informatics (for conception of the system and application software) have to be mentioned for their contributions to project realization.

S. Coman, A. Mircea, V. Papadopol and V. Vlad from the Central Institute for Management and Informatics have to be specifically mentioned for their significant contributions to the concept of the system.

Finally, acknowledgements are made to the Central Institute for Management and Informatics for assistance and for permission to present this paper.

AN INTELLIGENT MEASURING STATION OPERATIONAL FOR FLOOD WARNING IN BELGIUM

A. Van der Beken*
J. Marivoet and L. Deriemaker

1. Introduction

Water problems all over the world, have changed the scope of hydrology; once belonging to the category of pure earth sciences, dealing with the presence, the distribution and the movement of water on earth, it has evolved into an important engineering science which is at the basis of large scaled studies and expensive hydraulic engineering works with important social-economic and environmental impact.

This evolution during the last decades has been enhanced by generalised use of the computer which allows for vast quantities of field-measurements to be treated and analysed. Computer models are now being used for all kinds of hydrological applications, whilst in earlier days one had to be satisfied with oversimplified empirical formulae.

The progress of hydrology, however, is strongly influenced by the availability and reliability of hydrological measurements, such as precipitation , temperature, pressure, sunshine, humidity, wind speed and direction, water levels, water velocities, etc. Water quality and its numerous parameters constitute a second group of variables for which continuous field measurements is still very unreliable.

The Laboratory of Hydrology of the Free University Brussels, in co-operation with the Centre for Statistics and Operations Research and the Laboratory of Informatics, has developed various research projects dealing with quantitative and qualitative aspects of water. Our attention was chronologically first drawn to the study of the automatic control of flood reservoirs. It was necessary to develop "intelligent measuring stations" (IMS) to pursue this goal.

2. The automatic control of a flood reservoir

By means of the example of the planned flood reservoir of 2,8 Mm^3 at the river Dijle at Neerijse, approximately 10 km upstream of Leuven (Louvain), the aim of these intelligent measuring stations will be demonstrated. Fig. 1 shows the catchment of the river Dijle and indicates the place where the proposed flood reservoir is to be realized. The purpose of this reservoir is to regulate the discharge of the Dijle in such way that the discharge capacity at Leuven will not be exceeded in order to avoid flooding. However, two tributary rivers, Voer and Molenbeek, are not controlled by the reservoir, which means that controlling the reservoir (filling and emptying it) has to bring into account the discharges of both tributary rivers. As these control measures are only noticed some time later at Leuven (roughly 2 to 3 hours), there is a control problem which can be dealt with by a process computer (Fig. 2).

This process computer uses online-measurements, hydrological models and a decision model to calculate the filling and emptying of the reservoir on a 20 minutes' time base (Fig. 3). The process computer

*Professor of Hydrology, Laboratory of Hydrology, Free University Brussels (VUB)

World Meteorological Organization (ed.), Microprocessors in Operational Hydrology, 97-112.

proceeds to giving the necessary instructions with regard to the movable reservoir's gate. A simulation of such an automatic control during a storm is given in Fig. 4.

The online-measurements feeding the process computer are collected at seven stations in the downstream part of the river Dijle catchment (Fig. 5). Water levels, rainfalls and temperatures are measured on a 20 minutes' time base. The whole configuration (process computer and measuring stations) is represented in Fig. 6. More details are given in various publications of the Laboratory of Hydrology (see references at the end of this contribution).

3. The intelligent measuring station (IMS)

An efficient real time control of flood reservoirs is only feasible if the process computer disposes of recent measurements of river water levels and a number of meteorological variables at various places in the river catchment. To collect this information and to transmit it to the process computer an intelligent measuring station is necessary at each measuring station. Each IMS will control at least two of the following measuring instruments :

- limnimeter : type pressure sensor; an electrical signal (4-20mA or voltage) proportional to the pressure on the measuring membrane is available
- pluviometer : type tapping bucket; after 0,2 mm rainfall the bucket tips and causes a short-circuit in a mercury switch
- thermometer : type platina resistance thermometer (Pt 100); a signal conditioner gives an output signal (4 - 20 mA)
- relative hygrometer : type dry-wet bulb thermometer; consists of two resistance thermometers
- wind speed meter : generates electrical pulses proportional to the wind speed
- wind direction meter : 8 wind directions are digitalised (3 bits)
- barometer : the output signal (4 - 20 mA) is a linear function of the atmospheric pressure
- solarimeter : generates a weak electrical signal proportional to the captured radiation energy; a signal conditioner amplifies this signal (4 to 20 mA).

The functions of the IMS can be divided in two groups : automatic and non-automatic. The automatic functions must be accomplished autonomously once in each measuring cycle (20 minutes in this example). The following automatic functions are necessary :

- real time clock
- carrying out a number of tests
- processing of the signals from the measuring instruments
- registering the measured values in the memory along with the time
- carrying out a number of tests concerning the collected measuring values

and, if necessary, warming the central computer
- generating a pulse for the "watch dog" circuit.

The non-automatic functions are as follows :

- transmission by telephone line (*) (Fig. 7) of the stored data to the central computer when ordered to do so by the latter and, if necessary, to adjust the local clock
- carrying out any measurement, after the introduction of the appropriate key word by means of a portable terminal, to allow for repair and maintenance.

4. Hardware of the intelligent measuring station

A prototype of an IMS which can perform the functions listed above has been developed. The scheme of this IMS is represented in Fig. 8. It consists of modules, each of them being available on a printed circuit board, so that it can easily be adapted to local situations. The IMS has the following hardware components :

1° The bus

The bus connects all printed circuit boards with each other. It contains the following lines :
- power supply (5V, ± 12 V, earth wire)
- 8 data lines
- 8 address lines
- 4 interrupt lines
- 3 control lines
- 12 "chip select" lines
- 2 RS232 lines (Tx and Rx)

2° The processor board

A Z8 μ -processor with integrated timers and UART is used. One of the timers gives every 10 ms a pulse for the real time clock. An automatic starter mechanism resets the measuring station after cutdown in power supply. This board has a maximum memory capacity of 2K byte EPROM and 24 K byte static RAM.

(*) In other applications radiotransmission may be more indicated than telephone lines for connecting the measuring stations with the central computer.

3° The digital input board
....

Two types of digital input boards have been foreseen. A first type has a parallel input of 16 bits. The second type has three 16 bits counters with galvanic separation of the inputs and the system by opto-couplers.

4° The analog input board
....

Ten analog signals can be switched to the analog to digital convertor (ADC). An integrating ADC is to be preferred in order to filter out possible noise on the input signals. This ADC has a measuring range of $\pm$ 2 V and a resolution of 1 mV.

5° The signal conditioner board
....

This card transforms the various analog signals (tension, supply, resistance) to $\pm$ 2 V signals to make a precise analog-digital transformation possible.

6° The modem interface board
....

This board generates the missing modem signals so the UART of the Z8 μ-processor can be connected with a modem or, alternatively, with a pocket terminal. Furthermore the board is provided with 5 relays.

7° The modem board
....

Up to now there is no homologated modem that can be built in. For the ultimate version of the IMS it would be preferable if the modem could be built in, in order to obtain a compact and robust field station and to avoid further homologation problems with the Telephone and Telegraphe Administration.

8° An interchangeable memory board
....

Because of its modular conception, the IMS can also be used as a data logger for applications in which telemetry connection between the station and the central computer is not necessary. For that purpose an interchangeable memory board must be added to the system. This board can be supplied either with EPROM or RAM memory in CMOS technology with battery supply on the board. This board is replaced periodically and taken to the central computer where the stored data can be read.

5. Realisations

A first proto-type of the IMS is operational since March 1983 and controls the meteorological station at the campus Oefenplein of the Free University Brussels (V.U.B.). The following instruments are collecting data on a 10-min time base :

- 2 type of raingauges
- 1 air-thermometer
- 1 hygrometer
- 1 barometer
- 1 solarimeter
- 1 wind speed meter
- 1 wind direction meter

A typical example of hourly averages is given in Table 1, generated by the laboratory mini-computer Digital MINC PDP-11. Transmission of the collected data is automatic once a day. A similar table format for the 10-min data can be commanded once a day or at any moment.

These data are stored on floppy discs for about one month. Afterwards they are transferred to the mainframe computer for further processing.

The reliability of this system during the period 1.1.83-1.6.84 was about 80 %. The missing data were the result of surcharge of the Minc computer and the fact that the IMS was a prototype.

The nominal total hardware cost of this prototype, exclusive the measuring instruments, the telemetry system and the mini-computer, can be rated at ca. 60.000 BF.

Following the experience with the prototype, an adapted new printed board version has been developed. Hence, reproduction of several IMS is now possible at a nominal hardware cost of ca. 40.000 BF. One IMS is already installed at the storm drainage outlet and detention basin of the campus. Measurement of velocity and water levels are controlled. Several more IMS in the storm sewer system will be installed in the near future in the frame of an interuniversity research project on design and management of storm sewers.

The first pilot measuring station for the "Dijle real-time flood reservoir system" has been installed since January 1984 at Korbeek-Dijle about 1 km downstream of the planned reservoir. Until now, only the water level in the river and the rain is measured on a 10-min time base and continuously transmitted to the laboratory mini-computer by a leased telephone line. A typical example of the graphic output is given in Fig. 9, generated by the mini-computer Minc PDP-11.

Recently the NCR Tower 1632 mini-computer was installed. Its storage capacity (Winchester disk) is about 15 Mbyte for data. In the near future a link will be established between the mini-computer and the mainframe computer (CDC Cyber 170/750). The gathered data of the mini-computer will then be transmitted daily to the mainframe computer for back-up reasons and for further processing.

The reliability of this station during the short operational period was 70 % of the time. The missing data were the result of a break-down of one of the modems with long delay of repair (17 %) and surcharge of the MINC-computer (13 %). It is expected that the new installed NCR-Tower 1632 will avoid this surcharge in the future and that reliability will increase above 95 %. Total nominal costs for hardware of the IMS (exclusive all other elements) is estimated for this station at ca. 60.000 BF.

The short experience in this field proved that the operational procedures are extremely important in order to cut down the delay of repairs. Both hardware and software design are very time consuming. The benefit, however, is that the developped system is a turn-key system at a very fair price (research and design not included).

6. Future applications

By the end of 1984, the following IMS should be installed :

1. 4 IMS in the Dijle-basin and 1 IMS in the Demer-basin - connected with the minicomputers of the Laboratory of Hydrology through dialed telephone lines.
2. 4 IMS on the sewer system of the V.U.B. campus- connected with the same minicomputers through a dialed telephone line.

In the near future, other field measurement apparatus will be equipped with off-line IMS-hardware, used as dataloggers with replaceable EPROM-memory boards.

ACKNOWLEDGMENT

The authors would like to thank the authorities who supported the study financially and all persons who contributed by their help.

REFERENCES

ANONYMOUS (1982), Studie en programmering voor een real time computerbeheer van een hoogwaterreservoir op de Dijle te Neerijse (Study and programming for a real time computermanagement of a flood reservoir on the river Dijle at Neerijse (Guidebook), Final Report, Free University Brussels (VUB), Brussels, Belgium; 189 p.

BAUWENS W., MARIEN J., MARIVOET J.L., VANDEWIELE G.L. & VAN DER BEKEN A. (1983), The real time control system of the reservoir on the river Dijle (Proceedings of the HS4 Symposium, IAHS at IUGG, Hamburg) IAHS Publ. (in press).

MARIEN J. (1980), Regulation of a flood reservoir with the use of on-line forecasts, Hydrological forecasting (Proceedings of the Oxford Symposium) IAHS Publ. no. 129, pp. 477-483.

MARIEN J. (1982), Over de dimensionering en sturing van hoogwaterreservoirs (About the design and regulation of flood reservoirs), Doctoral thesis, Faculty of Applied Sciences, Free University Brussels (VUB).

MARIVOET J.L. & VANDEWIELE G.L. (1980), A real time rainfall-runoff model, Hydrological forecasting (Proceedings of the Oxford Symposium), IAHS Publ. no. 129, pp. 409-418.

VAN DER BEKEN A., VANDEWIELE G.L., TIBERGHIEN J., MARIEN J. & MEERSSEMAN M. (1980), On-line flow forecasting for automatic operation of a flood reservoir, Hydrological forecasting (Proceedings of the Oxford Symposium), IAHS Publ. no. 129, pp. 557-564.

LIST OF FIGURES

LIST OF TABLES

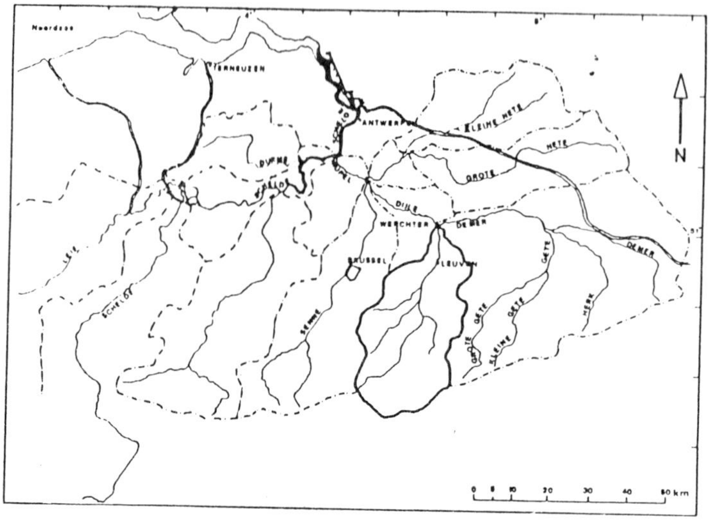

A. The Dijle River basin as part of the Scheldt River basin.

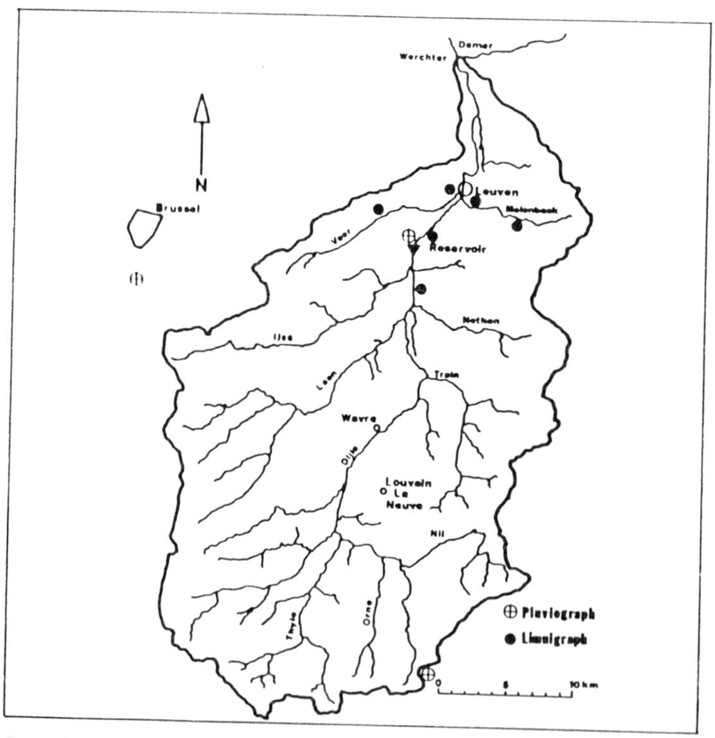

B. The projected flood reservoir with existing measurement stations

Fig.1: The catchment of the river Dijle

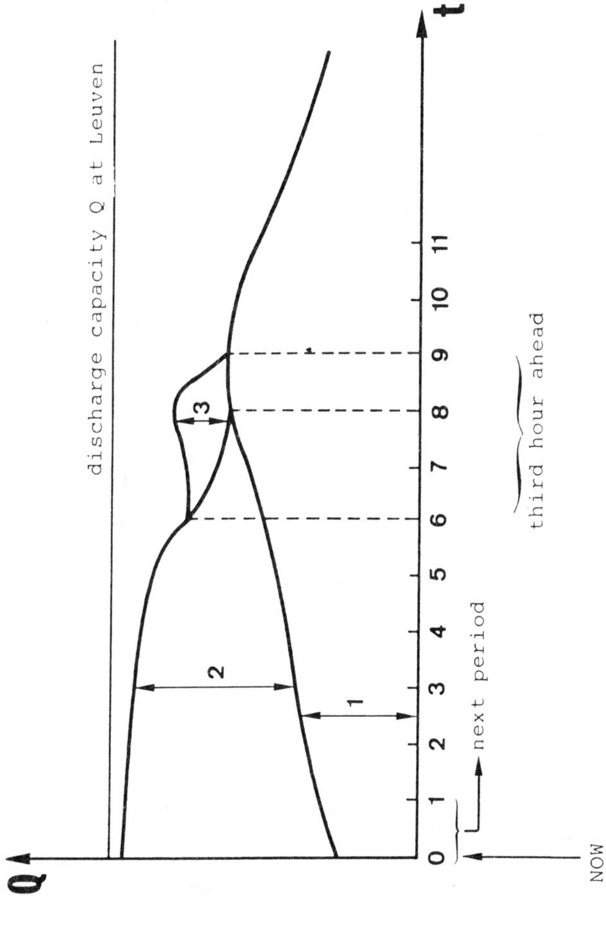

1: Discharge of Voer and Molenbeek
2: past reservoir discharges
3: present reservoir discharge

Fig. 2: The control problem of the flood reservoir

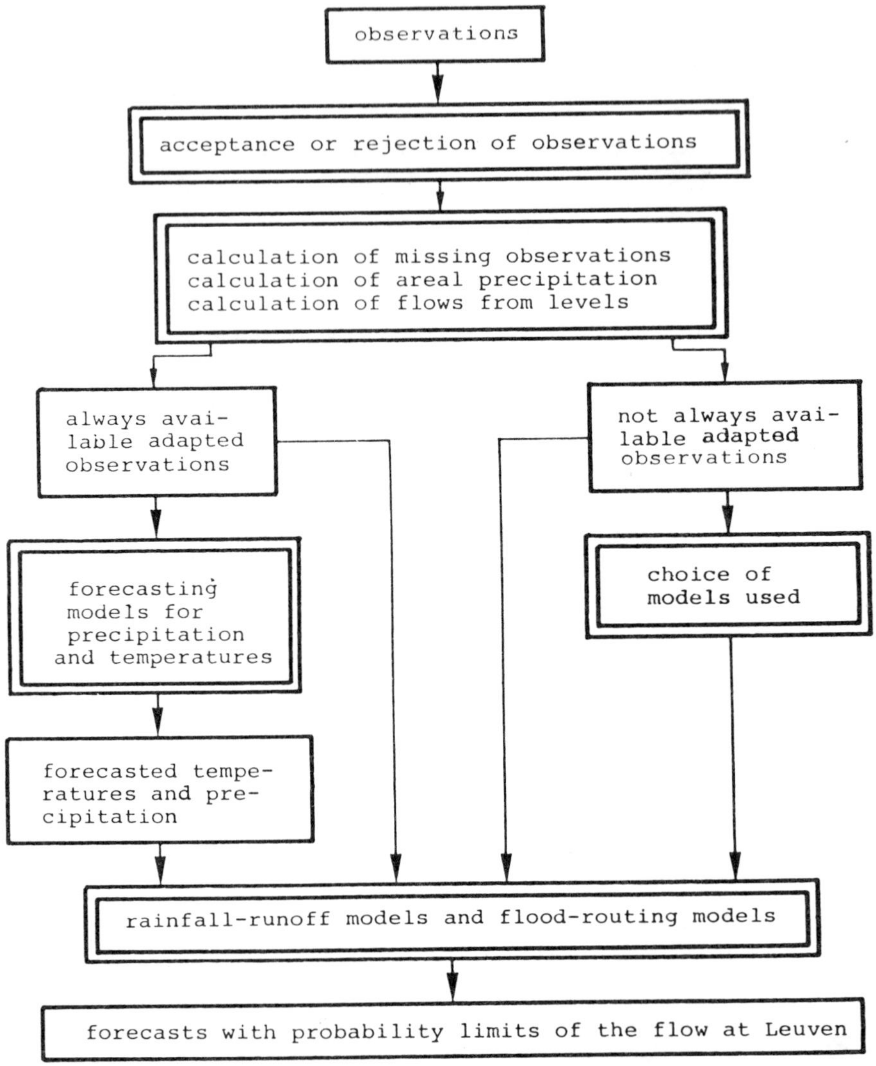

Fig 3 : On-line calculations from the observations to the forecast

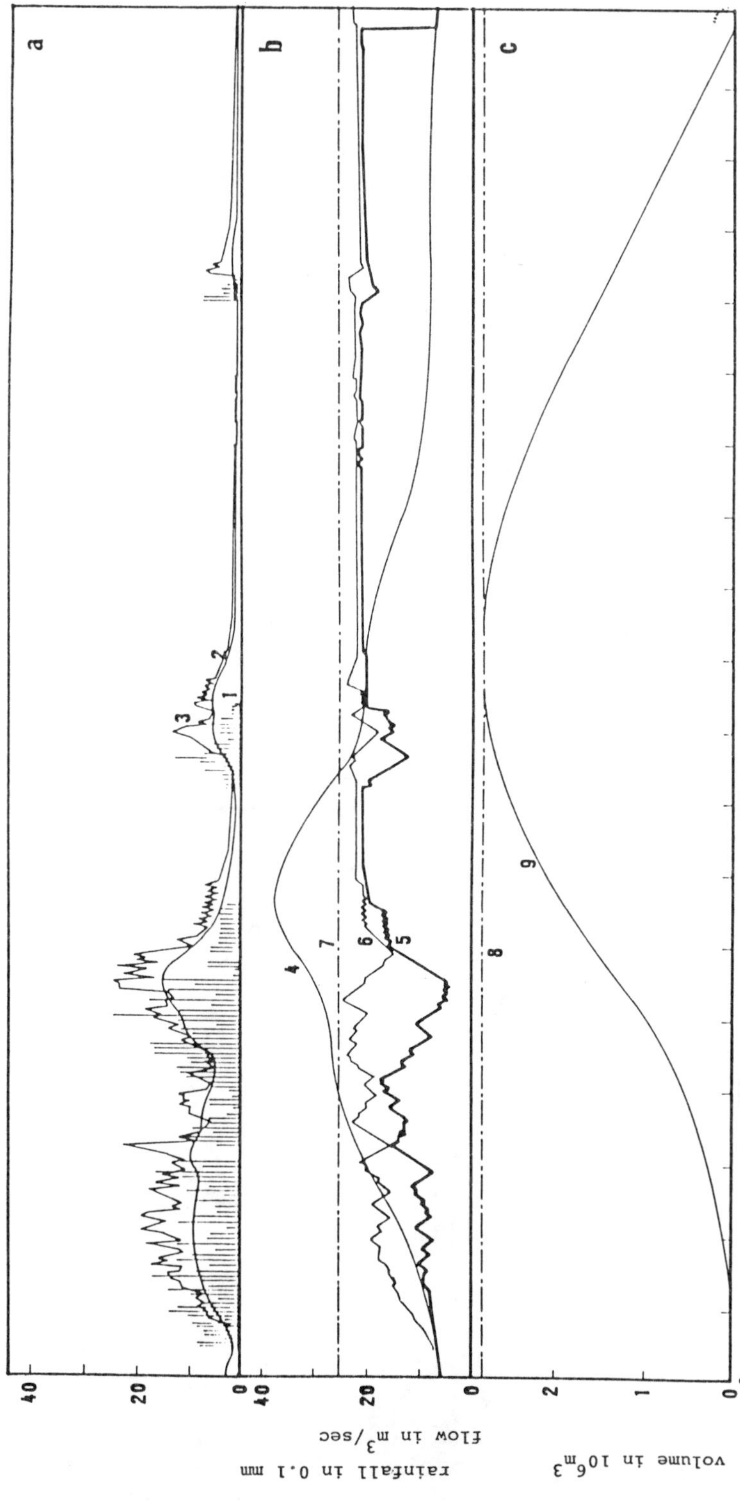

Fig. 4. An example of flood control (time in periods of 20 minutes)

a. curve 1: rainstorm in 0.1 mm
curve 2: quick response of uncontrolled area
curve 3: forecast uncontrolled flow

b. curve 4: inflow into the reservoir
curve 5: releases from the reservoir
curve 6: total discharge at Leuven
curve 7: critical discharge at Leuven

c. curve 8: maximum reservoir capacity
curve 9: reservoir filling

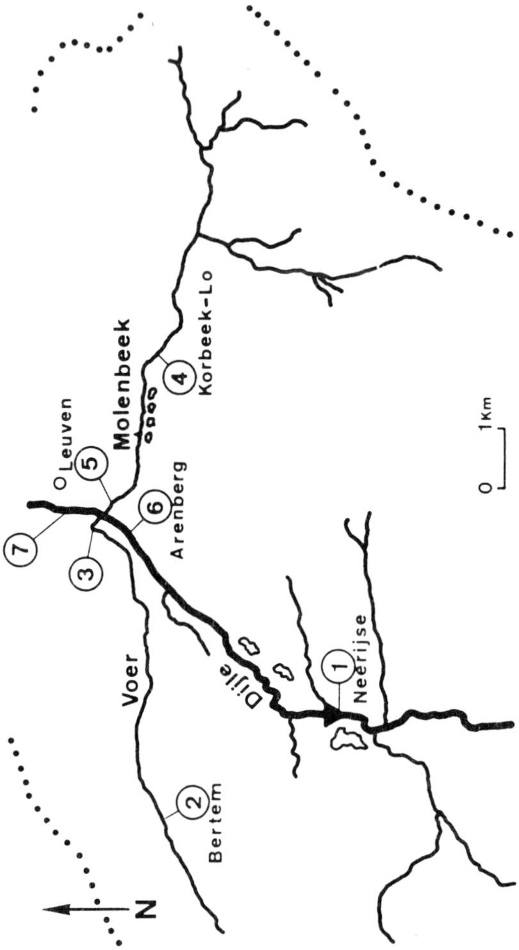

Fig 5: Location of the measuring stations

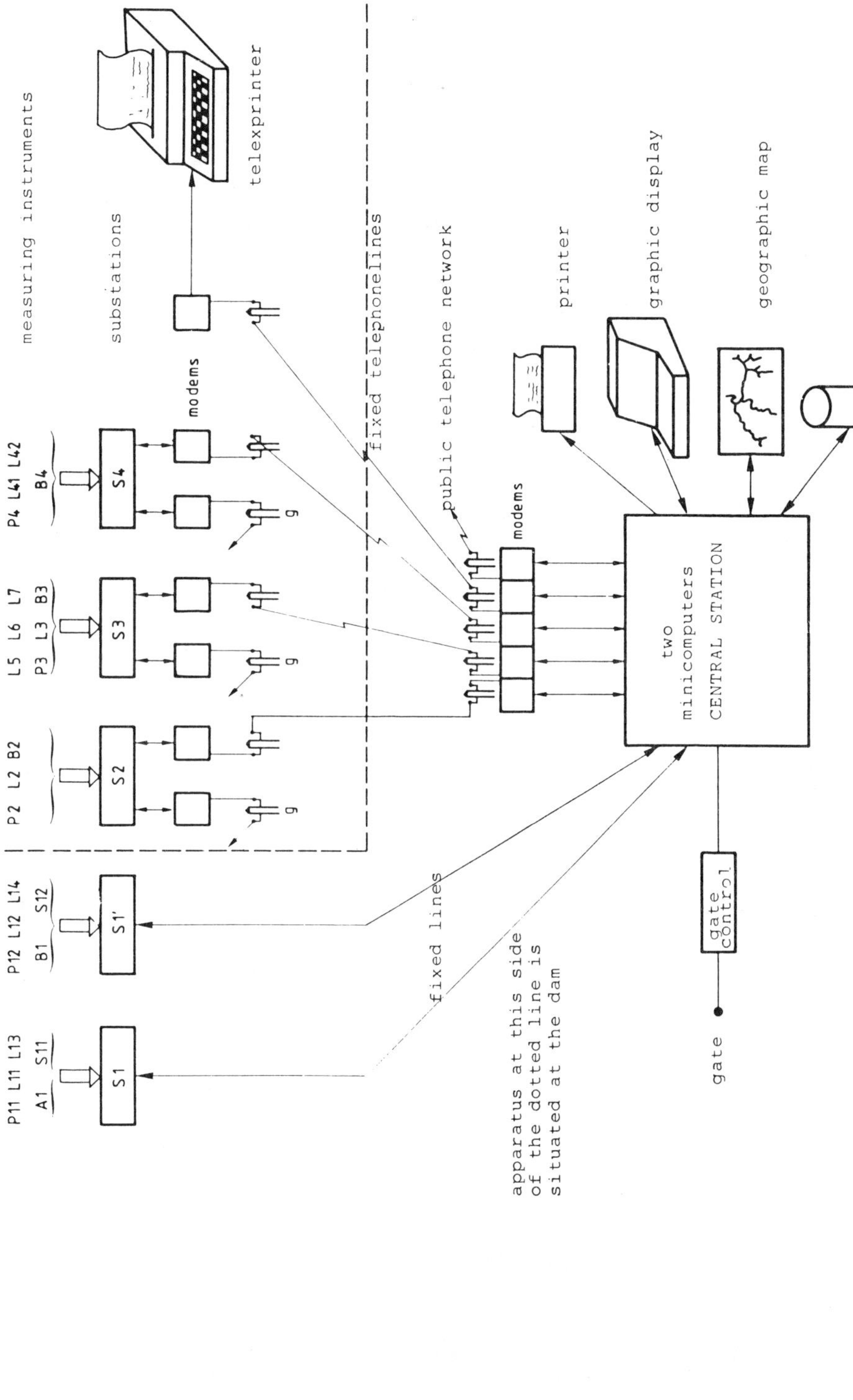

Fig. 6: Scheme of the links between the measuring stations and the central station

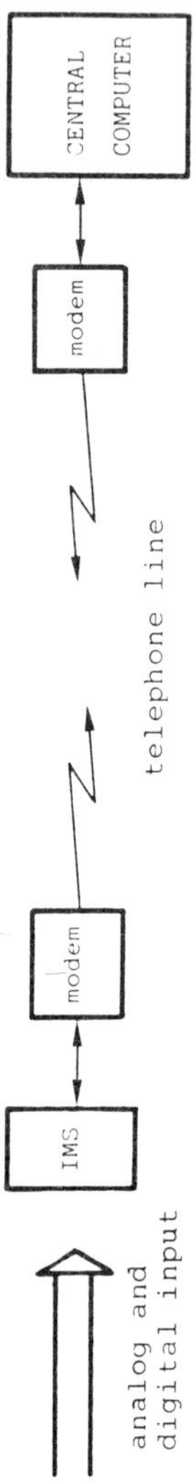

Fig 7: Scheme of the telemetry system between measuring station and central computer

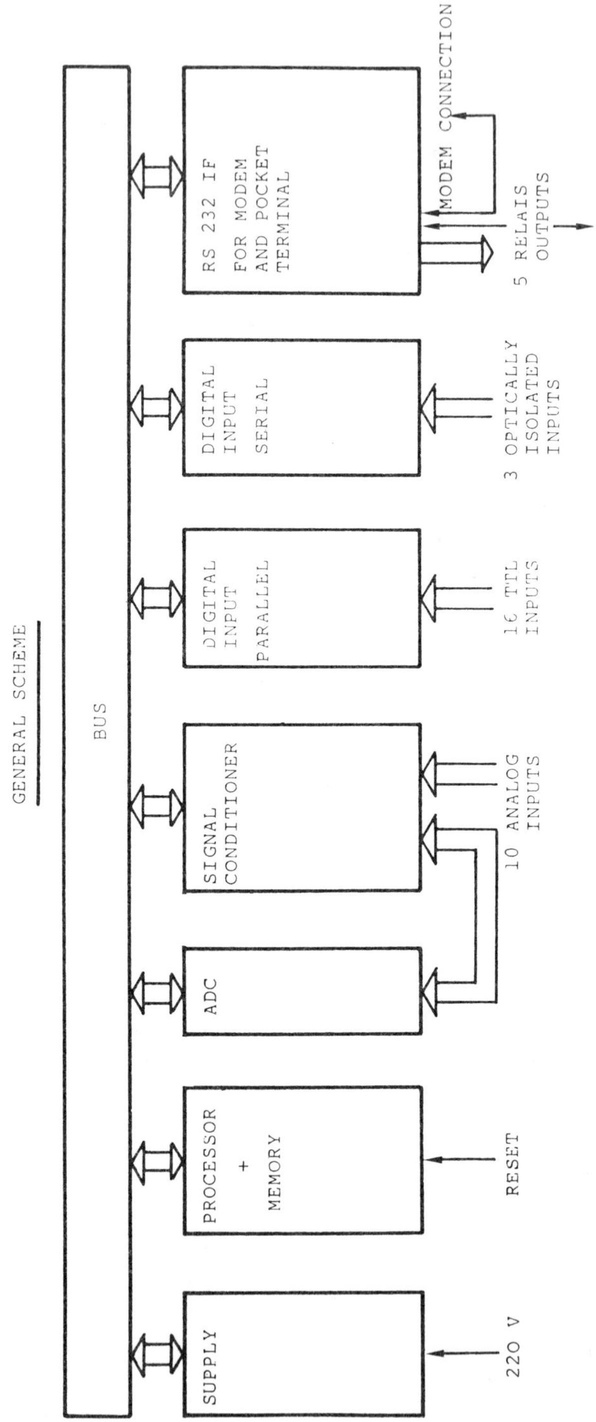

Fig 8: Scheme of the hardware of the intelligent measuring station

RAINFALL

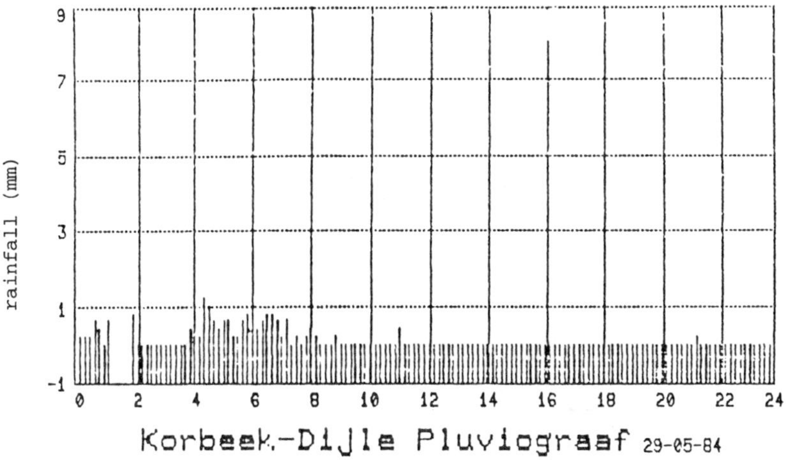

Total precipitation: 15,6 mm (resolution: 0.2 mm)

STAGE - DISCHARGE

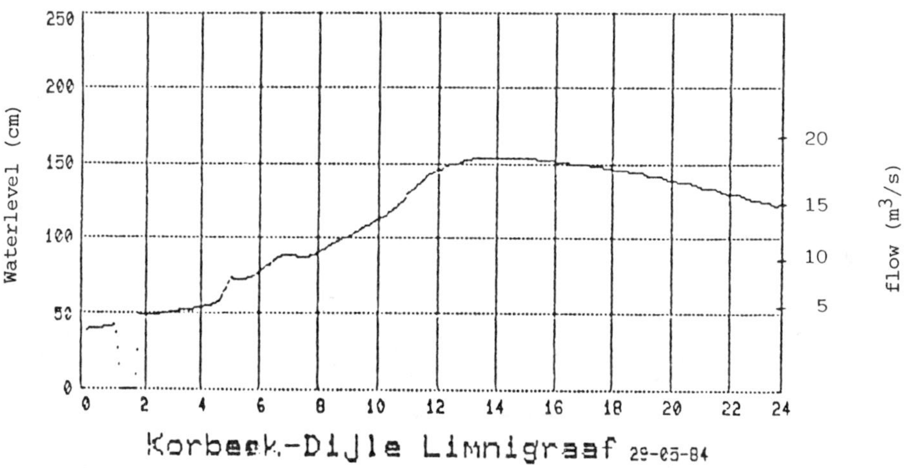

(resolution: 1 cm)

Fig. 9: Graphic output of the pilot measuring station at Korbeek-Dijle on May 29th, 1984 (10 min. time base)

6-12-83 HYDROLOGISCH MEETSTATION 1 - OEFENPLEIN

TIJD	PLUV.1 mm	PLUV.2 mm	THERM. *C	HYGRO. % (+)	BAROM. mbar	SOLAR. W/m*m	WISNEL Km/u	WIRICH
00u00	0.2		3	84	1013		6	ZW
01u00	0.4	0.5	2	94	1013		4	ZW
02u00		0.5*	2*	100*	1013*		5*	ZW*
03u00	0.3	1.0	2	103	1013		3	ZW
04u00			3	104	1013		3	W
05u00	0.4		4	105	1013		2	W
06u00			4	105	1013		2	W
07u00			4	106	1014		1	W
08u00			4	105	1014	5	1	W
09u00	0.2		4	105	1015	39	3	W
10u00			5	102	1016	110	2	W
11u00			6	98	1016	126	4	W
12u00			6	102	1016	109	4	W
13u00	0.2		6	100	1016	107	4	W
14u00			6	99	1016	38	5	W
15u00			6	98	1016	40	6	W
16u00	0.4	0.5	5	98	1017	4	7	NW
17u00			5	96	1017		6	W
18u00			5	96	1017		5	W
19u00			5	94	1018		5	NW
20u00			4	96	1018		5	NW
21u00			4	98	1019		4	W
22u00			4	99	1019		4	W
23u00			3	99	1019		3	W
MAX.	0.4	0.5	7	107	1019	228	8	W
MIN.			2	81	1013			
GEMID.			4	99	1016	24	4	
TOTAAL	2.6	2.5						

(+) Incorrect observations * incomplete average

procent geldige waarnemingen : 98.61 : percentage of meaningful observations

Table 1. Hourly averages of measured meteorological variables at the university campus on December 6, 1983

PRIMARY PROCESSING OF HYDROLOGICAL DATA ON MICROCOMPUTERS

K. Futaki[x] and Z. Vargay[xx]

In the early Seventies a new race of computers appeared on the world market, called the microcomputers, which presented the following advantages vs. the "classical" type of equipment, namely:

- no need for air conditioned rooms
- no need for specially trained staff
- easy installation in User's offices

However, these machines, at start have shown some disadvantages as well:

- limited memory and storage capacity
- fairly long access time

These hinderances have, as a result of constant developings, almost fully disappeared, enabling the microcomputers to solve more and more complex tasks.

The recongnition of the above mentionned advantages of the microcomputers has incited the Hydrological and Water Resources Department of the WMO to create, in collaboration with the hungarian Research Centre of Water Resources /VITUKI/, a program package for primary processing and storage of hydrological data applied to microcomputers.

The used Data Base Management System has been developed by VITUKI in FORTRAN programming language.

The program sequence has been adapted first to a TANDY TRS-80 Model-II, having TRSDOS operating system, and later converted for the standard CP/M operating system /version 2.2/.

The use of the sequence does not need special training, the programs are elaborated for technicians level. The installation of the sequence can be performed within 4 to 6 weeks included the training of the staff.

---

x Head of Computer Centre, Research Centre of Water Resources /VITUKI/, Budapest, Hungary

xx Research associate, Research Centre of Water Resources /VITUKI/ Budapest, Hungary

World Meteorological Organization (ed.), Microprocessors in Operational Hydrology, 113-125.

DATA IDENTIFIERS ON THE H D B

During the execution of the I/O-operations a set of hydrological data is handled. The location these data originate, the time of observation and the hydrological phenomena are coded in the data identifiers accompanying the data records. The general form of data identifiers is the following:

A - B - C , where

A contains information about the location
B contains information about the time of observation
C designates the phenomena itself which the data is relating to

Thus, the used data identifier is divided into three keys:

A station code
B year
C data type code

RESTRICTIONS IN USING DATA IDENTIFIERS

- due to the used DBMS, the record identifier may contain only numeric keys
- in principle the sequantial order of the keys is optional but for a given installation of the data processing system this sequence is fixed in the programs
- the lengths of the keys are fixed as well, and their values are the following:

  . station code 1 to 6 digits
  . year 4 digits
  . data type code 2 digits

- the length of the station code seems so to be variable but after initializing the data processing system, this length is fixed by the User - for this reason leading zero are to be used in the station code, if applicable
- the data type codes are also fixed in the different data processing programs, in order to enable them to recognize the necessary physical files. These codes can be changed only if the DATA statement is modified in the relevant program. The general form of data type code is the following:

XY, where

X indicates the data type
Y contains additional information about the data

X = 1 precipitation data /in tenth of millimetres/
= 2 water level /in centimetres/
= 3 water discharge /in m3/s/
= 4 suspended sediment discharge /in kg/s/
= 5 bed load discharge /in kg/s/

Y = 0 or 5 measured data
= 1 or 6 converted data /using relationships/
= 2 or 7 corrected data /using the continuous correction scheme/

Examples: the code 20 designates the measured, i.e. observed water level on current file
the code 57 designates the corrected bed load discharge data on archive file

- the relationships used in the data processing system are coded in another way. The general form is the following:

XiXj, where

Xi and Xj indicate the type of data participating in the relationship
Xi designates the type of the independent variable
Xj designates the type of the dependent variable

Examples: the code 23 designates the water level vs. water discharge rating curves
the code 34 designates the water discharge vs. suspended sediment discharge rating curves

STRUCTURE OF THE HYDROLOGICAL DATA BASE

The structure of the HDB, i.e. the interconnection between physical files is shown in Fig.1. The Station file is in a particular position, and this particularity should be kept even in case of application of the data processing system on larger computers. The Station file is the "main entrence" of the HDB. Each data processing program working with station codes checks first to establish, whether the requested station is valid or not for the HDB /i.e. it existis on the Station file/. The request is rejected until the station is initialized on the file.

The data processing programs are connected to the HDB through a buffer of transit zone. Thus, the operations resulting any kind of modification on the HDB enter first the new information into this transit zone, and only if the User decides by using special command /*W/, those can be transferred to their final place on the HDB. The transit zone, depending on its size, either is placed in the memory, or it may be the Work file itself as shown in the figure. The length of the zone should be enough for one station characteristic record, or a rating curve, or a whole station-year record.

MAINTENANCE OF THE H D B FILES

During the evolution of the Data Base, files have to be opened, copied, dumped or the User may be interested in the current state of the files. The program HDBMNT performs all these tasks. After starting it prompts the User, by screening an * /asterisk/ to enter command.

Available program commands are the following:

*I Initialization of HDB files.

A HDB file has always to be initialized before using the DBMS subprograms on the file. The DBMS needs special information about the file during the execution of the I/O operations. Under the control of this command the User can enter these information into the file Header Record. If only a reset to zero is needed for a file /e.g. Work file/, a $ /dollar/ should be entered when the program asks for the file key limits. Otherwise, the LOWER and UPPER key limits are to be specified using any kind of separator /e.g. blank/ between the key subfields of the data indentifiers.

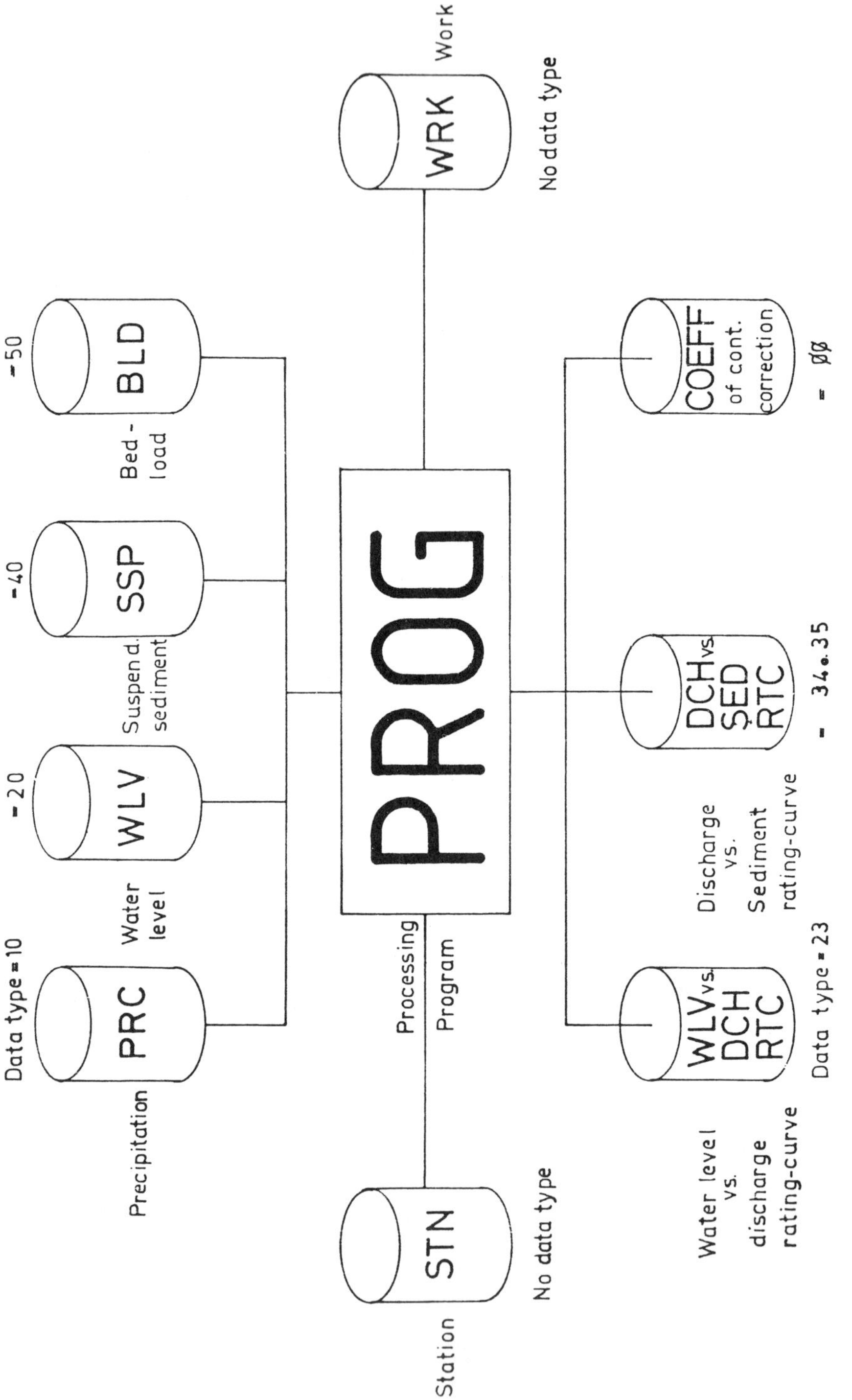

Figure 1. Physical files of the Hydrological Data Base

*X Extension of an existing HDB file size.

The write operation always checks, whether the requested data record /and the associated address records/ can or not be written on the HDB. If there is not enough place on the HDB file, the subprogram does not start the write operation but it stops the execution with an ICB/5/ = 6 error /i.e. the file needs extension/. This extension can be performed with this command.

*L Listing of current contents of the Header Record.

The use of this command provides a list of current contents of the file Header Record. This list can be a very useful help for the User to have an idea about the state of the HDB file.

*D Dump of HDB file, sector by sector.

Usually this operation is not needed in normal use of a HDB file. However, having the possibility of the file dump is indispensable while debugging the Data Base Management System.

*C Copying of a HDB file content to another.

This operation transfers each data record /within the defined limits/ from the Source HDB file to the Destination HDB file. Before copying, the Destination file should be initialized similarly to the Source file. This operation transfers only the accessible data records from the Source file, thus, the dummy area will be ignored /CLEAN operation/.

*F Quick summary of a HDB file content.

The use of this command provides a summary table for the HDB file. The present of a data record is marked with an * /asterisk/. This command should not be used for the Station file!

*Q End of program execution.

*E End of program execution.

STRUCTURE OF THE H D B FILES

The internal structure of a HDB file is created by the DBMS, which is a combined hierarchical-pointer system. The advantages of the DBMS are the following:

- it keeps the Data Base in the most compact form
- it handles data records of variable length
- it does not need sorting

Since reading is the most frequent operation executed on the HDB, the DBMS subprograms are provided with special features to optimise,before all, the performance of the read operation.

In the file system of the HDB, it existis two sets of files namely:

- Current file, containing the station-year records of the current year. The data records are stored on this type of file in uncompressed form because of frequent overw riting
- Archive file, containing the same station-year records in compressed form. Once the data processing of the current year is terminated /i.e. all data of the year are stored on the file/, the data records should be compressed using the HDBCPY program. This program transfers the data records to the Archive file in compacted for.

The designer of the present system has developed such data compression methods, which are able to compress the different type of data records to the 25 to 30 percent of their initial length.

Since the data processing programs and the DBMS subprograms are written in FORTRAN, it is more convenient to store INTEGER data values into the HDB, however, in case of discharge values /water~, suspended sediment~ or bed load~/, the type of data is typically REAL. To avoid this storage problem, an algorithm of data conversion can be used, which results in coding the converted data value /rounded to the third valid digit/ and its exponent in an INTEGER variable.

PROGRAMS OF THE DATA PROCESSING AND STORAGE SYSTEM

The programs of the system work in interactive mode which is the most convenient with microcomputers. The logical scheme of the data processing system /i.e. connection between the HDB and the programs/ is shown in Fig. 2. The basic unit of the communication is the message. The general form of the message is the following:

part A - part B - part C - part D, where

part A defines the type of the message
. blank ordinary, informative message
. ? program question
. ! error message
part B the text of the message in English
part C the possible User's answers between parenthesis
part D the default answer between brackets
This answer can be generated by a simple press on ⟨CR⟩ key.

Note: in a concrete message one or several of these subfields may be missing.

The programs check the User's answer to the maximum extent and in case of error, an error message is screened before repeating the question. The User, at any level of the communication, has the possibility to interrupt by pressing $ /dollar/. The User's answer is always informal, i.e. the answer may be started at any position of the line, and in case of entering more than one data in one answer, an arbitrary number of spaces /or any other separator/ can be used in between. There is only one restriction: the answer should be terminated in the same line.

The programs offer always two output devices, either the screen or the line printer. The line printer should be able to print at least 80 characters per line and the use of USA character set is suggested.

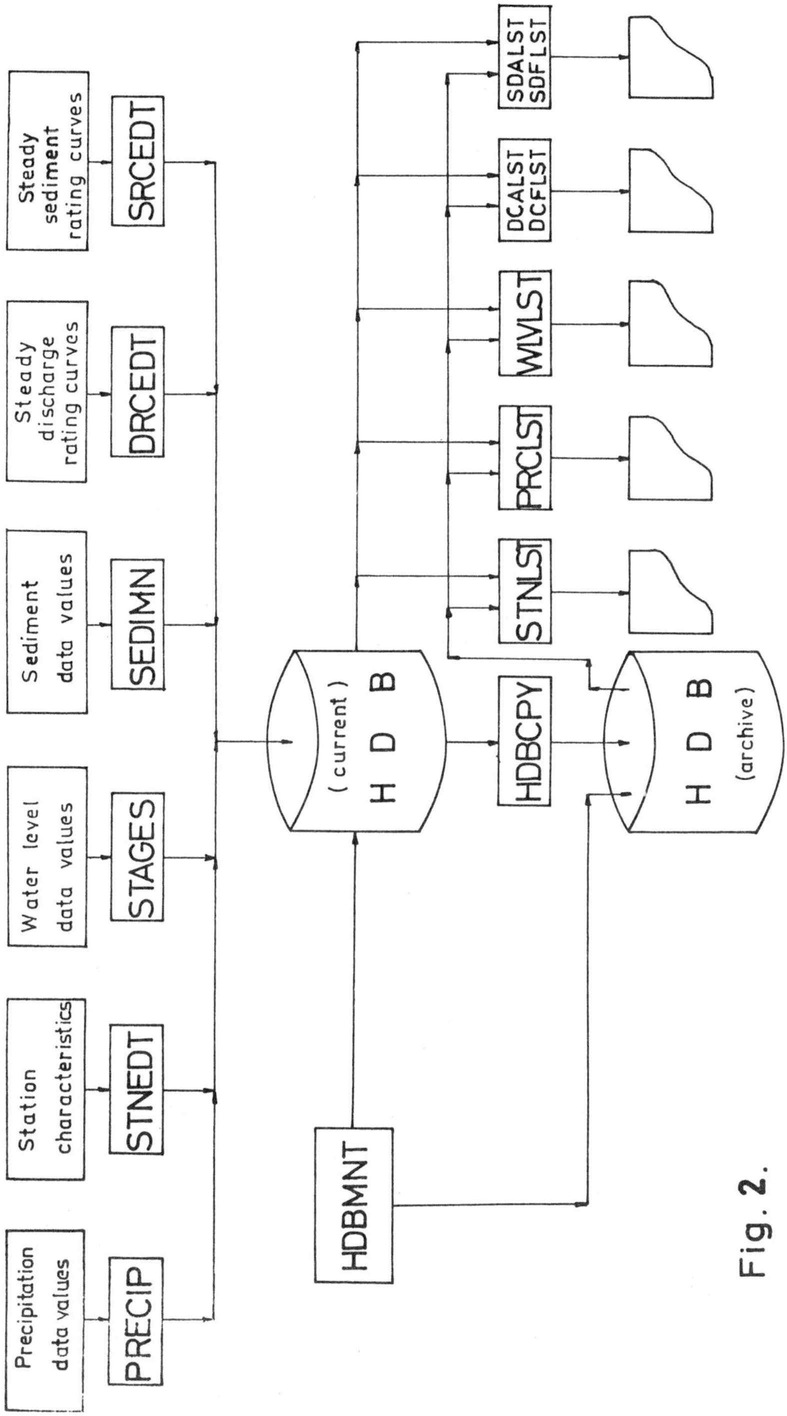

Fig. 2.

Logical scheme of the hydrological data processing and storage system

UTILITY PROGRAMS

The utility programs of the data processing and storage system are, in one hand, to make easier the usage of the system, and on the other hand, to do all hauskeeping work on the HDB.

During the whole work the User is aided by the program HDBHLP. This program, after entering the program name, screens a short description of the requested program task. The needed file/s/ and the possible program commands with their meaning.

After starting, HDBHLP asks for the program name. At this level by simple pressing the ⟨CR⟩ key, a list of the available program names is screened. To exit the program, a "Q" should be entered instead of the program name.

The I/O operations are executed by the DBMS /the TRN-subprograms/. If a transfer error is detected by the DBMS, the relevant subprogram transmits an error code to the calling program which decides, whether an immediate abort is needed, or the execution may be continued. In case of aborting, the calling program screens the current content of the transfer control block /ICB/ and the data identifier /KEYS/, underlining the error code on the screen. The program TRNROR gives the textual meaning of these error codes.

The program HDBMNT has been explained earlier.

The HDBCPY program transfers the data records from a current to an archive file, or if it is necessary /e.g. data correction is needed on an archive file/ from archive to current file. In addition HDBCPY is able to execute transfer operations between current to current or archive to archive files.

After starting, the program asks for the transfer conditions, where each step is checked and confirmed by the program, then the transfer is performed. The transfer phase is terminated when all station-year record is transmitted to the destination file within the specified limits. If no new transfer phase is requested the program stops.

EDITION PROGRAMS

The programs of this group handle the information necessary for the use of the HDB. These programs do not check the entered data, their main task is to update the relevant files.

The programs prompt the User by screening an * /asterisk/ to enter the desired command. A subset of the program commands is commonly used by the edition programs namely:

*W Writing of the current contents of the program transit zone into the HDB.

*E Exit program with writing.

This command writes the current contents of the program transit zone into the HDB, and stops the execution.

*Q Exit program without writing.

This command results in exiting the program, without transferring the current contents of the program transit zone into the HDB.

Station record handling program

The program handles the STATION file contents, i.e. the basic characteristics of the stations. After starting, the program prompts the User by screening an * /asterisk/ to enter command.

The available program commands are the following:

*I Insertion of a new station into the Station file.
*M Modifying of the basic characteristics of a station.
*D Deletion of a station /record/ from the HDB.
*W, *E and *Q see above!

Rating curve handling programs

These programs enter that discharge rating curves which were elaborated manually by the User. The programs are able to handle only steady rating curves /i.e. only two-parameter-relationships/, which are given by min. 2 and max. 33 data couples. The range of values differs for the different data types:

- precipitation data	from	0.0	to	999.9	mm
- water level	from	-979	to	9999	cm
- discharges	from	0.0	to	10.	to the 22nd power m^3/s or kg/s

The data couples should be entered in increasing order, and the first resp. the last data couple will be considered as the range of the relationship. After entering all data couples, a SPLINE function will be fitted to the entered points and the curvatures of the function will be stored with the rating curve points on the HDB.

Each rating curve has one /or several/validity periods which are stored in the rating curve directory in increasing order. The rating curve directory is divided into "pages" on the HDB, where each page consists of a fixed number of lines /i.e. validity periods/. Interpolation between two validity periods is possible when using the appropriate flag value in the RTCFUN subprogram.

The programs accept only one rating curve for a given period. Derivation of rating curves from one to another is not allowed in the programs, in such case the whole derivated raing curve should be entered.

The program DRCEDT handles the water discharge rating curves elaborated by the User. The rating curve points are to be entered in form of water level /in centimetres/ and water discharge /in m^3/s/ data couples. The program is able to accept up to 99 different rating curves for each station. The highest number of the directory pages is 99 and one directory page contains 7 lines, thus, the total number of the validity periods is 693 for one station.

The available program commands are the following:

*I   Insertion of a new rating curve.
*M   Modifying of the rating curve directory line/s/.
*D   Deletion of an existing validity period from the directory.
*L   Listing of the rating curve directory or the tabulated rating curves.
*W, *E and *Q see above!

The program SRCEDT handles the sediment discharge rating curves elaborated by the User. The rating curve points are to be entered in form of water discharge /in m^3/s/ and suspended sediment or bed load discharge /in kg/s/ data couples. The program is able to accept up to 99 different rating curves for each station. The highest number of the directory pages is 99 and one directory page contains 5 lines, thus, the total number of the validity periods is 495 for one station.

The available program commands have the same meaning as in DRCEDT program above.

DATA PROCESSING AND STORAGE PROGRAMS

The programs of this gorup execute the main task of the system: the data processing and storage. Different programs exist to process different type of data, and the structure of these programs is very similar. The modules of the programs have always the same task but the algorithms, specially those which execute the data checking, are different, depending on the type of data to be checked. The checking procedures are the following:

- comparison with a given range
  . for every type of data
- comparison of the consecutive data differences with a given range
  . for water levels
- comparison of the slopes/upper and lower/ with a given range
  . for water levels.

The data checking programs prompt the User when a violation of the current checking limit is detected. The detected error is screened then with a note on the type of the error and the programs block the screen in order to prevent the User to continue entering data. In such case the programs offer two possibilities to the User:

- enter a correct data value without changing the current checking limit/s/
- to change one or several checking parameters in order to make accept the entered data value.

Both of these acts can be easily executed. If the User "opens" the checking limit/s/ for a data, there is an easy way to reset the original limit values, and continue the data processing with the more strict limits.

The programs of primary data processing need at least one data value for a day, and they are able to process maximum 96 data /within 15 minutes/. If more than one data exist for a day, the User is allowed to choose an appropriate time step code in order to enter the data value in the appropriate moment. The available time step codes are the following:

- one data per day
- one data per hour
- one data per 30 minutes
- one data per 15 minutes
- no automatical time incrementation.

In case of using the first four time steps, the programs increment automatically the time, according to the current time step code, the User has just to enter the data at the appropriate moment. Using the last time step code, the correctime of the observation should be specified with the data value with the accuracy of 15 minutes. In this case the change of days can be done by simple pressing the ⟨CR⟩ key.

The programs use common code for the missing data: -999, which has a different meaning of the code of initial value: -998. This latter means that the User has not yet entered any data for the given moment /or day/.

The data processing programs allow the User to enter a supplementary information with each data value. The information is stored in one byte, and may be a hydrological phenomena code or a qualification code defined by the User. The present data processing system is able to recognize:

- precipitation phenomena code /with the precipitation data values/
- ice phenomena code /with the water levels/
- qualification code /with the sediment discharge data values/.

The meaning of these supplementary information depends on the User's needs, and they can be easily changed in another system installation.

The programs, similarly to the edition programs, prompt the User by screening an * /asterisk/ to enter the desired command. The write / *W/ and the program termination / *E, *Q/ commands have the same meaning as for the edition programs.

<u>Processing program of precipitation data values</u>

The program PRECIP processes and stores the precipitation data values. The data values should be entered in tenth of millimetres followed by the appropriate phenomena code. The phenomena codes are the following:

blank	for rain
*	for snow
X	for sleet
:	for granular snow
%	for hail
&	for freezing rain
#	for thunderstorm

In case of having rain in trace the code -100 should be used.

The available program commands are the following:

*I Insertion of new precipitation data values.
*M Modifying existing precipitation data values.

Processing program of water levels

The program STAGES processes and stores the water level observation values. The data values should be entered in centimetres followed by the appropriate ice phenomena code. The ice phenomena codes are the following:

blank for no ice
> for ice on the board
: for drifing ice
I for standing ice
for ice jam.

The available program commands are the following:

*I Insertion of new water level observation values.
*M Modifying existing water level observation values.

Sediment data processing programs

The program SEDIMN processes and stores the discharge values either for the suspended sediment observation or for the bed loads. The entered values should be computed first manually /or with the help of a simple program/ and for both types of data the average discharge value of the cross--section should be entered. The unit of the entered data value is kg/s for both types of sediment data, and it should be followed by the appropriate qualification code. The available qualification codes are the following:

blank for measured data
C for computed data /with relationship/
E for estimated data /the data definition method is different of the previous one/

The available program commands are the following:

*I Insertion of new sediment discharge values.
*M Modifying existing sediment discharge values.

LISTING PROGRAMS

The listing programs are able to produce uniform lists of the processed data values. These lists contain always only the data values previously entered into the HDB with the exception of the water discharges. These are converted from the water levels using the available water discharge rating curves. In case of the edition programs the listing may be produced by the programs themselves /see DRCEDT or SRCEDT programs/ by using the appropriate program command / *L/, or a separate program produces the listing, e.g. STNLST program for the basic characteristics of the stations.

The listing programs all operate in a similar way:

- a range of data identifiers may be defined for the selection operation.
- the User decides for the type of the listing.

  . A if annual data tables are to be produced either with daily mean values or with the observations relating to the specified time
  . M if monthly and annual extreme values are to be listed
  . D if detailed input data values are to be listed /this is valid only if more than one data were specified for a given day/
  . F if frequence and annual duration tables are to be produced /not valid for precipitation data values/.

All listing program uses identical symbols in the following cases:

- the User has not yet entered any data for the given moment: ...
- inexistant day: blank
- the code of missing data was entered for the given moment: ---

In case of water discharge listing the sequence "?RC" may appear if any error of the used rating curve/s/ is detected.

The code of rain in trace is converted in 0.0 on the listing of precipiation while blank is listed for the days having no rain.

If once during the whole year the code -999 appears on the station-year record the programs do not produce the monthly and annual extreme values. The presence of -998 for a whole day results the same effect. The request of frequency table / .F subcommand/ is ignored in such case.

In case of frequency and duration table request, the interval range can be defined either by the User /water or sediment discharge listing/ or may be fixed in the program /10 cm for water levels/.

If the value to be listed is greater than 100.000 its value appears on the listing in the following form:

DDDsE, where

DDD is the first three valid digits of the data value
s is the sign of the exponent
E is the exponent

e.g.: 367-2 = 367 10**-2=367 0.01 = 3.67
367+5 = 367 10**+5=367 100000 = 36700000.

The listing programs recognize both the current and the archive files.

x x x

The authors thank the co-workers of the WMO Hydrological and Water Resources Department for their efficient help extended to them during the elaboration of the data processing program sequence presented hereby. They also wish to express their gratitude towards the South-East Asian representative of the WMO for his invaluable contribution to the work executed.

MICROTIDEDA

M.E.U. Taylor[1] S.M. Thompson[2] M.W. Rodgers[3]

INTRODUCTION

microTIDEDA is a computer system for processing time-dependent data, particularly hydrological data. The software, like other data base systems can be used to collect, edit, store, display and analyse the information it is given. microTIDEDA uses programming techniques which allow numerical data to be filed in the minimum amount of disk space and retrieved at high speed. This has enabled the transfer to microcomputers of software which originally required a mainframe computer to run.

microTIDEDA deals with data that are naturally recorded as a time sequence. Examples are stream flows, rainfall, air temperatures and electric power generation. The lack of an elaborate data structure in such data sequences allows simple processing methods, and only the large quantity of data justifies the use of an electronic computer. The microTIDEDA program accepts data from a variety of automatic recording instruments. The data is stored in a form designed to facilitate checking, correcting and analysis, and a comprehensive set of aids is provided for these purposes.

When microTIDEDA is manipulating data, it can apply a combination of linear and non-linear transformations. The linear transformation allows the application of scale and additive factors, whilst the non-linear transformation is performed by applying a user-defined function to the data. This function is entered as a series of discrete values, and for hydrological data allows the representation of stage-to-discharge rating curves.

Features available in microTIDEDA fall into five categories:

1 There are processes which provide general statistical summaries. These include plots, print-plots, and tabulation of means, extremes and totals computed for daily, weekly, monthly, seasonal or annual periods.

2 There are processes which allow data entry. These accept numbers from either a 16 track paper-tape reader, a keyboard, or a suitably formatted disk file.

3 A number of processes deal with the correction, storage and deletion of data, the manipulation of files and batches, and the efficient use of space on disks.

4 There is a process which lets the user write his own data analysis software using a simple programming language.

5 There are several processes which control and software's operating environment. These specify the disk files to be processed and enable the user to carry out a succession of pre-defined tasks which can then be run without further investigation.

1 Research Director, Water and Soil Directorate, Ministry of Works and Development Head Office, P.O. Box 12-041, Wellington North, New Zealand

2 Scientist in Charge, Hydrology Centre, Ministry of Works and Development, P.O. Box 1479, Christchurch, New Zealand

3 Scientist, Hydrology Centre, Ministry of Works and Development, P.O. Box 1479, Christchurch, New Zealand

World Meteorological Organization (ed.), Microprocessors in Operational Hydrology, 127-139.

HARDWARE

To use microTIDEDA, an 8-bit microcomputer which supports CP/M is required, together with a line printer and A3 plotter. If 16-track paper tapes are to be handled, an appropriate paper-tape reader will also be needed. The New Zealand Ministry of Works and Development (MWD) uses microTIDEDA on standardised equipment, and each field party is issued with an Altos microcomputer, a VDU, line printer, A3 plotter and 16-track paper tape reader. The Altos was chosen because it is cheap and has improved performance using an AMD maths processor. microTIDEDA is also available in IBM personal computer, DEC Rainbow and DEC VAC versions.

SOFTWARE

microTIDEDA is designed to be user friendly. It provides a menu of alternative options covering the various hydrological data processing options and assists the user to decide what he wants.

PROCESSES

Each facility available in microTIDEDA is called a PROCESS. These fall into five general categories:

1 Processes which add data to a data disk.

2 Processes which display data from a data disk.

3 Processes which edit data.

4 Process PSIM which provides a programming language to analyse the data.

5 Processes which control the operating environment.

A pictorial view of the ways processes can manipulate the data appears in Figure 1.

After the name of a process has been entered, the program will prompt for the essential parameters which define the site and period of the data to be processed. Each parameter is displayed in a particular area on the screen called a 'field'. The values last used are listed for the users convenience, and the values chosen are displayed under the present settings. These can be overwritten or left unchanged. It is not always necessary to type all the parameters, and this may be simplified by using the SPECIAL KEYS.

OPTIONS

Options allow the user to modify the operation of a process. After the site parameters are entered, the available options are displayed together with their current settings. The value of each option is preserved from one process to the next, and there are no default values. This allows the same options to be used without re-typing them.

When an option is changed, the new setting overwrites the old value on the screen, and what is seen on the screen is what the process will use when it runs.

Some options have a numerical value associated with them. Others are only on or off. A few options take numerical values which can be turned on and off, e.g. when maxima or minima are set on various process options.

There are two groups of commands which do not act on the data but merely control the way microTIDEDA operates. One group allows the microTIDEDA disk files to be named. A second controls the 'learn' mode of operation. This permits a series of tasks to be set up for microTIDEDA which can be run without further intervention.

microTIDEDA ORGANISATION (FILES, BATCHES AND ITEMS)

microTIDEDA uses a single data file into which it puts all the data it is given. The file has an index at the front of it called the 'directory', and microTIDEDA uses the directory to tell where particular data will be found.

Although raw data is entered as discrete values, these represent a continuous series through time. The time may be unequally spaced; the more frequent the values, the closer the representation of the continuous process. A BATCH is a piece of contiguous data and the user can see the batch structure of the file using process BSCAN. This lists every batch put onto the file in the sequence it was entered.

A period between batches for which there is no data is called a GAP. In some cases gaps are marked in the data because of the data entry method, although there is no actual discontinuity. The user can delete these, and join batches together so that they form a larger contiguous period of data.

A quantity filed as a time series becomes an ITEM. Often only a single item is recorded for each time, but microTIDEDA can store up to 15 items provided all the items are recorded at the same time. Each item is identified by a number and this allows individual items to be manipulated without altering the rest of the data.

microTIDEDA DIRECTORY

Whenever a batch is written into the file an entry is written into the microTIDEDA directory. If all of these data are later replaced by a single batch of new data, then the new data take precedence for subsequent processing simply because their entry is further down the directory.

However, when only part of an earlier batch is replaced by new data, several new entries are written into the directory; one for the new batch and others to describe the parts of the earlier batch that have been retained. The extra entries are called 'invisible entries' because they are not an original entry and so are not listed by process BSCAN.

For each invisible entry the earlier batch is read from disk and searched to find the location of the junction between the replaced and retained data. Subsequent processing can then take place by directly reading data in sequence using location information in the directory.

A helpful picture of the directory is obtained if each entry from a particular site is plotted on a time line across the page, with successive entries appearing in order down the page. When viewed from the bottom of the page visible entries, which correspond to new data, obscure any earlier data for the same time, but invisible entries are transparent and can be seen through to the earlier data.

There is a special directory entry called a 'deleting batch'. This indicates which section of the series data has been deleted. The user can subsequently retrieve the deleted data.

Processes which add data to a file automatically update the directory. During this updating the directory is linked, that is, the sequence of batches in time order is determined and held in memory. The new batch is compared with this sequence to determine whether any invisible entries are to be incorporated in the directory. Previous entries are not altered when the directory is updated and changes are made by the addition of new entries. Thus, the directory contains a record of all previous changes and a user can retrace his editing steps if required. When much editing is done the directory becomes large and the file fragmented.

INTERPOLATION OF SERIES DATA

When data is stored it is assigned a label which describes what KIND of data it is and therefore what form of interpolation is appropriate to estimate values between filed values. The three kinds of data are INSTANTANEOUS, HISTOGRAM and INCREMENTAL.

microTIDEDA takes advantage of the properties of these forms of data sets to minimise the storage requirements for time series data while ensuring that the original values can be exactly retrieved.

Figure 2 plots the same set of data which appear quite differently with the three kinds of interpolation. Values can be retrieved from each kind of data at times which need not coincide with the stored times, and either as instantaneous values or averages over an interval.

COMPRESSION

The removal of the intermediate values which can be deduced from their surroundings is called compression. Such values may be regarded as the middle of three which, when interpolated, form a straight line, thus:

Uncompressed

Compressed

The compression algorithm will allow a tolerance to be specified so that middle values within the tolerance of the interpolation are not stored. When a non-zero tolerance is specified for histogram or incremental data, the value of each element that is retained is adjusted to preserve the average or increment of the uncompressed data over the time interval it covers.

When retrieving data from a file the interpolation calculation is always invoked, and so there is no benefit by keeping uncompressed data on the TIDEDA file.

RATINGS

Recorded information must often be transformed by a rating or calibration curve. For example, ratings are used to transform water level records into stream discharges. Ratings may not be known when data are recorded, or may be changed later, or may change from time to time. For these reasons, each rating is stored as a separate batch and filed with the site number of the data with which it belongs, and the time from which it first applies.

Ratings are entered as a series of pairs of values (X,Y) and the X-value must always increase. A minimum of three and a maximum of 50 pairs may be entered.

Quadratic interpolation is made on the ratings by fitting a parabola through three points:

(Xi-1,Yi-1) (Xi-Yi) (Xi+1,Yi+1)

Where possible the three points are chosen so that X is between Xi-1 and Xi, and then the parabola is chosen to be either X on Y or Y on X, depending on which has the least curvature at the point of interpolation. When (Xi-Yi) is a maximum or minimum point, a Y on X parabola with its vertex at (Xi-Yi) is used. Extrapolation is carried out, where necessary, with the choice between a Y on X or X on Y parabola, again determined by which has less curvature at the point of extrapolation.

Often a rating change can be considered as occurring instantaneously, in which case it is filed with the time at which it changed. Sometimes a rating change is gradual, from one stable form to another. During the transition period, the rating used to convert data lies somewhere between the new rating and its previous form. This situation is allowed for by associating two times with the new rating, thereby creating a smoothed change. The earlier time, called the 'start' time, is the one at which the change is considered to be complete. In the overlap interval between the ratings, a linear interpolation is applied between the ratings. Given that Yold corresponds to X in the old rating, and Ynew corresponds to X in the new rating, then the value of Y is given by:

Y = Yold + (Ynew-Yold)*(time-start_time)/(overlap_interval)

where the overlap_interval = effective_time - start_time.

Ratings also have an item number, and so each item of series data can have its own rating curve. By convention, a rating with a given item number is applied to the corresponding item of series data, but this need not be so. Moreover, it is quite acceptable to have more rating items than items of series data. This allows the possibility of applying successive ratings, for example a stage-to-discharge rating followed by a discharge-to-sediment load rating.

HYDROLOGICAL CALCULATIONS

All conventional hydrological calculations, and tabular and graphical presentations can be quickly and easily performed using microTIDEDA. Two processes are worthy of particular note:

1 PSIM

A set of mathematical statements representing a physical system can be used on the stored data to simulate a process, with up to 15 input records from a file and up to 15 output records stored elsewhere in the same file. The language provides comprehensive facilities for this kind of simulation. The different records, which may be from different sources, may be read simultaneously and a record derived from them. This facility, used during data entry, can make good a missing period of record. Used at a later stage of processing it can analyse how changes in one component of a system will affect another, for example to predict how such developments as hydroelectric power, irrigation or other river control works will change the flow regime.

Other kinds of transformation are used to reduce the volume of stored data, including conversion of a seiching lake record to mean lake level, and separation of the tidal components from estuary records. In both these bases a record of inflow from upstream is then used for checking. Another kind of transformation converts records of head-water-level and gate-opening to flow.

A description of the effect of astronomical tides at a particular recorder site is stored in the form of 25 tidal constituents. This description, coupled with the stored residual time series, can be used to reconstitute the original record, and when making a flood forecast.

Applications of PSIM are limited only the ingenuity of the user.

2 QUALITY ASSURANCE CHECKS

microTIDEDA provides automated data quality assurance procedures. These can supplement or even largely supplant conventional quality assurance techniques.

The automated procedures can be used on quickflow or baseflow records and for stage-to-flow ratings. They assist earlier identification and rectification of faults. The quickflow test is carried out first. If a series record does not pass this first test the other tests need not be performed. The principle of the method is illustrated using data from three stations on the West Coast of the South Island of New Zealand: the Grey River at Dobson (401) and Waipuna (404) and on a tributary the Ahaura River at Gorge (407). The records are first separated into quickflow and baseflow (Figure 3). For each quickflow event identified, the quickflow volumes from each site are divided by the catchment area and give the quickflow depth of the event. The quickflow volumes of paired events from adjacent sites are plotted on a log/log graph using one site as ordinate, the other as abscissa. Quickflow errors in one site or the other usually appear as isolated values which plot well away from the bulk of the data (Figures 4, 5, 6).

The actual quality assurance procedure is more complex and rigorous than that described here and would usually be too arduous to carry out manually. It is easily applied routinely using microTIDEDA.

OPERATIONAL EXPERIENCE

microTIDEDA has been operating on minicomputers in each of the 16 MWD field party stations for some 1½ years now. Decentralisation of the data processing facility, which was previously run on a mainframe computer at head office, has shortened the time for production of verified edited data from a minimum of several weeks to an hour, or so. Editing can now be carried out while events are still fresh in staff's memory.

Job satisfaction has increased enormously. The system has been judged an unqualified success.

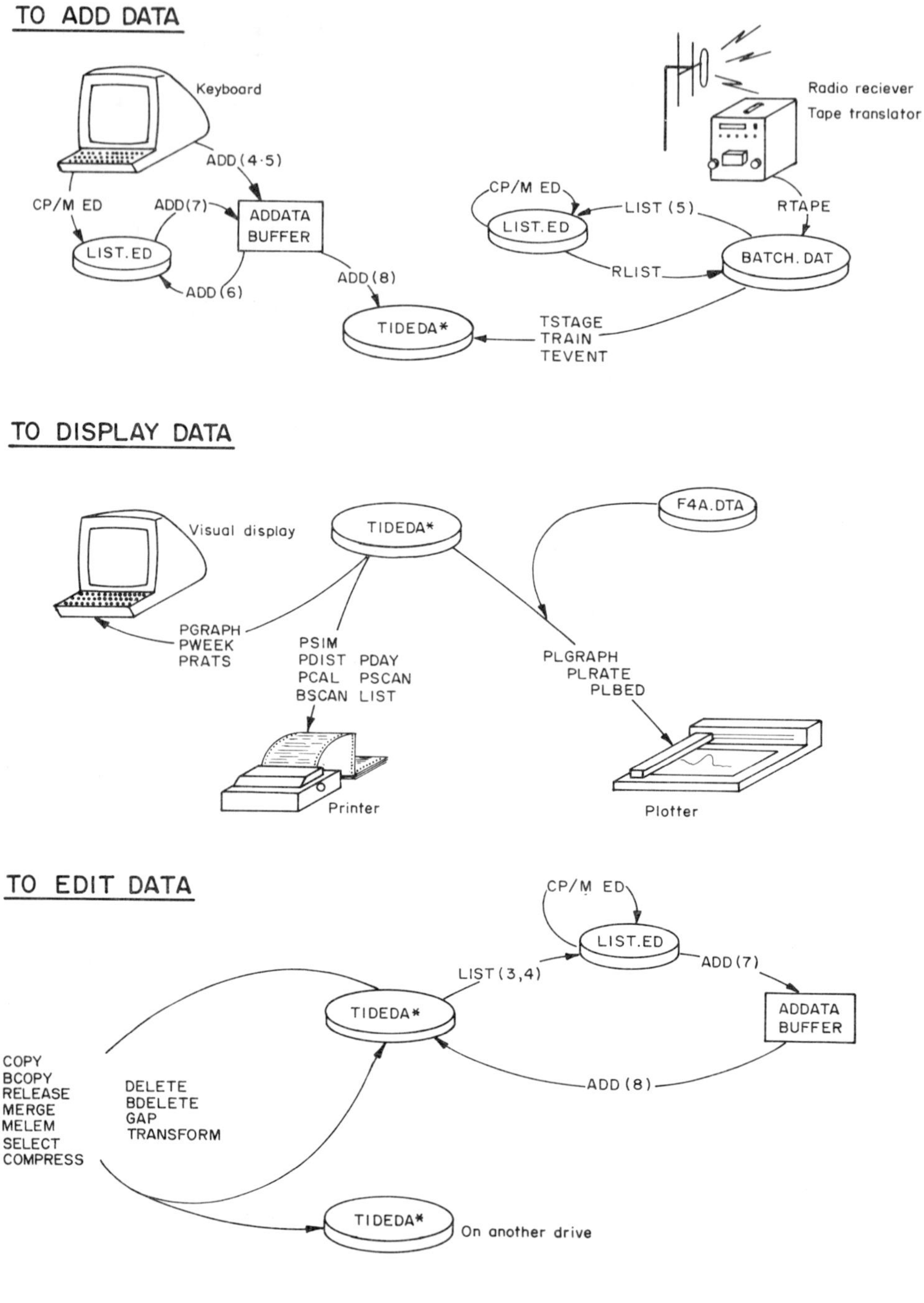

Figure 1: MicroTIDEDA Processes

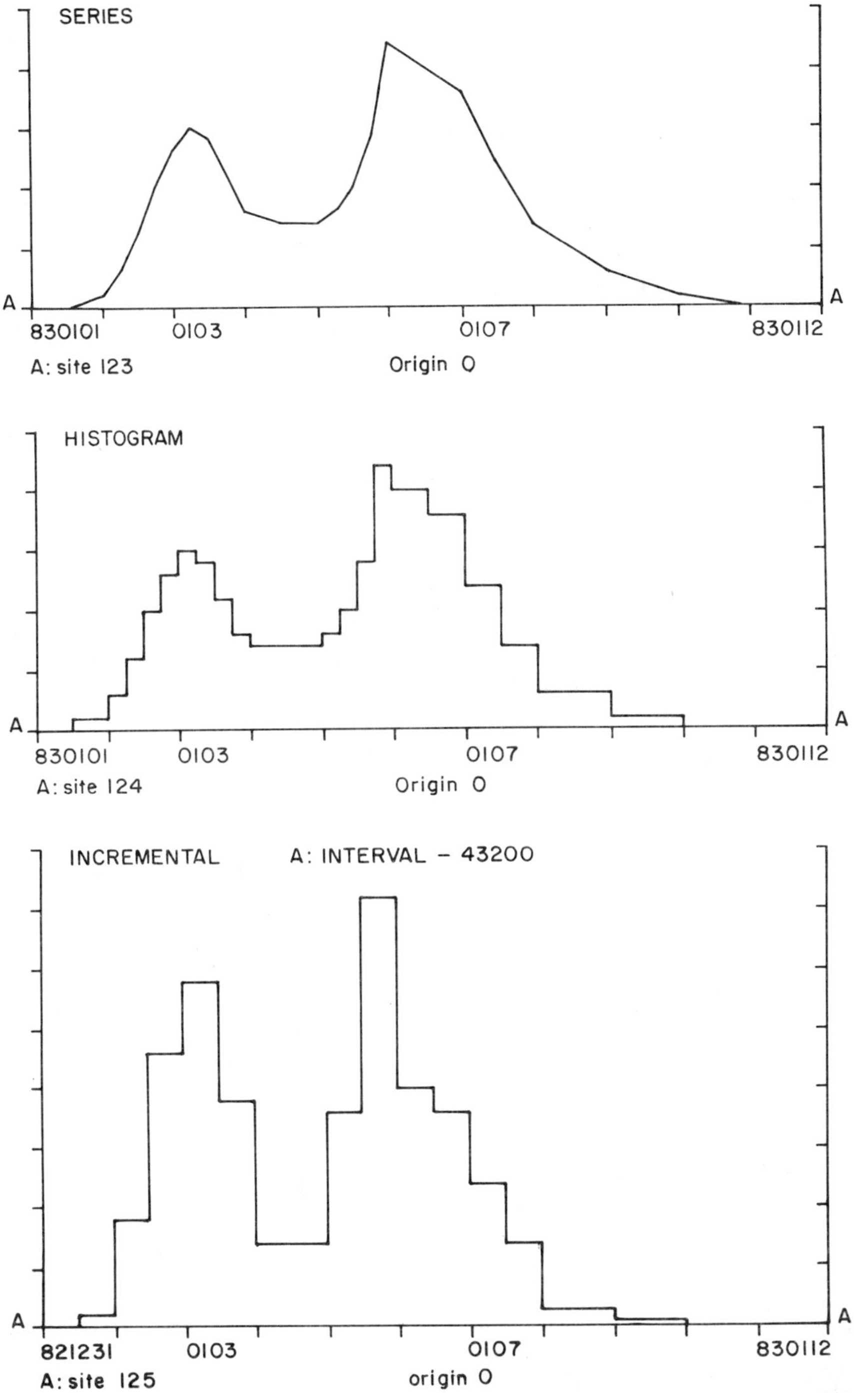

Figure 2: Kinds of data

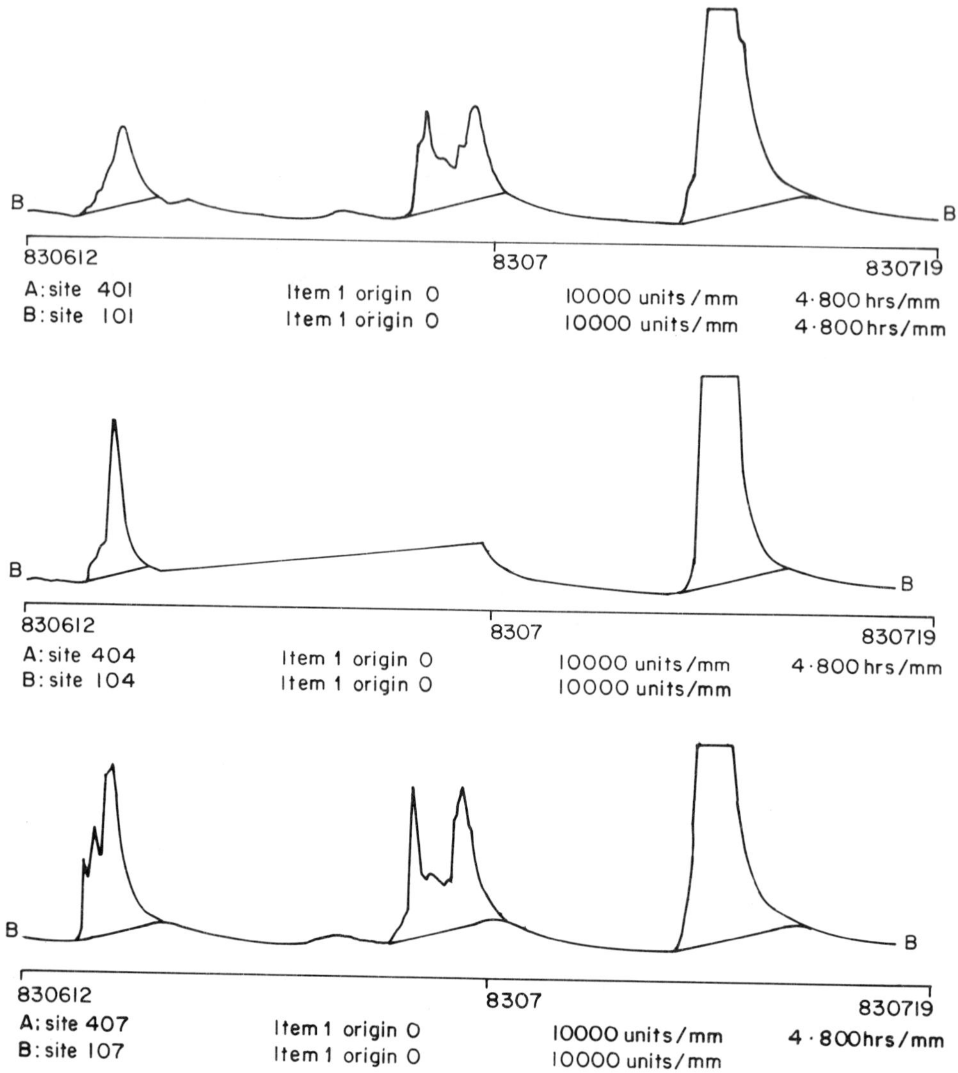

Figure 3: Records of flow per unit catchment area (units = pm/s) from recorders on the Grey River at Dobson (401), and at Waipuna (404), and on a tributary, the Ahaura River at Gorge (407). Baseflow separation lines are shown. The top of the last quickflow event has been truncated where it meets the plot frame.

```
          ### LATEST  EVENTS ###
37 events  from  830429 100000  to  830904 163000
   0 1 2 3 4 5 6 7 8 9 A

A                          A     1m

9                          9  300mm

8                  2       8  100mm

7    1           4         7   30mm

6              2 1         6   10mm

5                          5    3mm

4  1       1               4    1mm

3    1       1             3  300um

2                          2  100um

1  10  1 1                 1   30um

0    11                    0    0

   0 1 2 3 4 5 6 7 8 9 A
```

```
                 ###  ANOMALOUS  EVENTS  ###
7,1  4.25 days to 830701 114500  :  4,0  1.25 days to 830625 150000
4,4  1.51 days to 830611 104500  :  3,1  0.96 days to 830809 194500
3,5  1.05 days to 830501 234500  :  1,2  0.59 days to 830724 174500
1,3  0.69 days to 830814  43000
```

```
          ###  UPDATED EVENT TABLE  ###
63 events from 830101   1500 to 830904 163000
   0 1 2 3 4 5 6 7 8 9 A

A                          A     1m

9                          9  300mm

8                  4       8  100mm

7    1           6 1       7   30mm

6              5 2         6   10mm

5            1 2           5    3mm

4  1       1 2             4    1mm

3    1     1 1             3  300um

2  1       1               2  100um

1  13  1 1                 1   30um

0   15 1 1                 0    0

   0 1 2 3 4 5 6 7 8 9 A
```

Figure 4: Quickflow Depth Comparison No. 414
Ordinate : 91401 : Grey at Dobson
Abscissa : 91404 : Grey at Waipuna

LATEST EVENTS
29 events from 830429 50000 to 830904 163000

	0	1	2	3	4	5	6	7	8	9	A		
A												A	1m
9												9	300mm
8									2			8	100mm
7								5				7	30mm
6							2	1				6	10mm
5						1						5	3mm
4				1	1							4	1mm
3						2						3	300um
2												2	100um
1	10		1	1								1	30um
0		1		1								0	0
	0	1	2	3	4	5	6	7	8	9	A		

ANOMALOUS EVENTS
4,3 1.42 days to 830611 80000 : 4,4 1.66 days to 830625 150000
1,2 0.84 days to 830724 103000 : 1,3 0.69 days to 830814 24500

UPDATED EVENT TABLE
63 events from 830101 1500 to 830904 163000

	0	1	2	3	4	5	6	7	8	9	A		
A												A	1m
9												9	300mm
8									4			8	100mm
7								7	1			7	30mm
6							4	2				6	10mm
5						2	2					5	3mm
4				1	1	3						4	1mm
3						3						3	300um
2					2							2	100um
1	12		1	1	1							1	30um
0		11	2	2	1							0	0
	0	1	2	3	4	5	6	7	8	9	A		

Figure 5: Quickflow Depth Comparison No. 417
Ordinate : 91401 : Grey at Dobson
Abscissa : 92407 : Ahaura at Gorge

```
              ### LATEST  EVENTS ###
     39 events  from  830429  50000  to  831107 161500
            0  1  2  3  4  5  6  7  8  9  A

        A                                     A    1m

        9                              1      9  300mm

        8                           4         8  100mm

        7                     1  6  1         7   30mm

        6                     1  1            6   10mm

        5               1  1                  5    3mm

        4            2     1                  4    1mm

        3               1                     3  300um

        2         1  1     1                  2  100um

        1   10                    1           1   30um

        0      2     1  1                     0    0

            0  1  2  3  4  5  6  7  8  9  A
```

```
                ###  ANOMALOUS  EVENTS  ###
9,9  8.20 days to 830513 160000  :  7,6  2.83 days to 830904 170000
7,8  3.36 days to 830617 190000  :  5,4  0.86 days to 831101 214500
4,3  1.46 days to 830611 104500  :  4,3  1.18 days to 830919  51500
3,4  0.48 days to 830814  24500  :  2,2  0.68 days to 830920 184500
2,3  0.86 days to 830724 103000  :  2,5  1.08 days to 830809 191500
1,7  4.65 days to 830702   3000
```

```
            ###  UPDATED EVENT TABLE  ###
    75 events from 830101   1500 to 831107 161500
            0  1  2  3  4  5  6  7  8  9  A

        A                                     A    1m

        9                              1      9  300mm

        8                           8         8  100mm

        7                     1  9  1         7   30mm

        6                     2  3            6   10mm

        5               1  4                  5    3mm

        4            2  1  2                  4    1mm

        3   1           1                     3  300um

        2   1     1  1     1                  2  100um

        1   12    1               1           1   30um

        0      13 1  2  4                     0    0

            0  1  2  3  4  5  6  7  8  9  A
```

Figure 6 : Quickflow Depth Comarison No. 447
Ordinata : 91404 : Grey at Waipuna
Abscissa : 91407 : Ahaura at Gorge

EXPERIENCE OF MICROCPROCESSORS IN THE NATIONAL AUTOMATIC DATA ACQUISITION SYSTEM IN SWEDEN

Ture Hovberg and Jörgen Nilsson*

1. BACKGROUND

The Swedish Meteorological and Hydrological Institute (SMHI) is responsible for data collection in meteorology, hydrology and oceanography in Sweden. The Institute is organized to provide a common technical service including instruments, data acquisiton, communication and big computer operation. This organization facilitates the use of the same technical tools for different user purposes.

In 1969, the first two automatic weather observation stations were operating, and in 1975 the automatic station network contained five automatic stations with computer controlled data acquisition via the switched telephone network. These automatic weather stations were also soon used for experimental work with oceanographical and hydrological sensors, but technical modifications were proportionately difficult and expensive to make, since the automatic stations were built in "hardware" technology.

The 1975 purchase of new data acquisition stations revealed that the new stations would have to become more flexible in choice of sensors and data reduction algorithms, and this was also indicated in the new specification sheet. To fulfil this, the supplier chose a concept based on the INTEL micro processor 8080 and a simple software monitor, which facilitated use of the stations for different purposes. The first automatic station with micro processors was delivered in 1977, and also an INTEL MOS 800 workstation for software development.

At this time SMHI was involved in special studies concerning alternative sources of energy. The studies also contained an essential part of special measurements which would be possible to perform with the new micro processor automatic stations, if the software monitor was modified. While the delivery time from the supplier was too long for this software modification, SMHI made the decision to be independent of software consultants or instrument producers and began to develop its own micro processor software. Since this decision, SMHI has expanded the staff of micro processor software development, and invested in additional equipment for software production as well as staff training.

The fundamental work with the automatic data acquisition system was described by Hovberg 1980, see (1).

*Swedish Meteorological and Hydrological Institute, Nörrkoping, Sweden

World Meteorological Organization (ed.), Microprocessors in Operational Hydrology, 141-153.

2. CURRENT DEVELOPMENT

2.1 Networks

In 1984, seven years after the first microprocessor station was installed, SMHI operates 54 automatic stations (see fiqure 1), 52 stations with automatic data acquisition via the switched public telephone network, and 2 stations equipped with ARGOS transmitters. Thirty-four automatic stations are used in routine network for meteorological, oceanographical and hydrological data acquisition, and 20 in special networks for wind and sun energy measurements, see references (2) and (3). Another 40 meteorological automatic stations will be installed early in 1985. The main purpose of these stations is to provide surface information in the pilot system PROMIS 600 (Program for an Operational Meteorological Information System), where observations from the ground, satellite, and radar stations will be integrated into a system for short-range weather forecasting. See reference (4).

In order to decrease costs for the acquisition of observed information (mainly from manned synoptic stations) and to increase the data retrieval, the Institute has specified and ordered a new kind of equipment: micro computer based observer terminal - Man DATs - Man DAT is a small table terminal with keyboard, display, micro computer and power back-up, connected to the switched telephone network. The observer can use it for input of Synop data - coded or in parameter form - or hydrometeorological information. In SYNOP mode, an automatic error check is performed. Functions for calculations of air dewpoint and reduced pressure are included in the software. See reference (5).

2.2 System configuration

The Automatic Data Acquisition System, ADAS, is used for interrogation of Automatic Data Acquisition Terminals, ADATs. ADATs are the automatic stations used at unmanned meteorological, hydrological and oceanographical sites. The Automatic Data Acquisition Central, ADAC, is dimensioned both for the ManDATs and ADATs. The interrogation of ADATs and ManDATs is performed via the switched telephone network by means of a number of automatic calling units. The ADAC consists of two HP 1000 minicomputers, and has, besides communication, also short-time storage, presentation and plotting functions. ADAT is connected to SMHI's big central computer, UNIVAC 1100, where long time data storage and secondary data treatment is performed. One output from the ADAS is connected to the central message switching system ATESTO, which is connected to the WMO Global Telecommunication System, GTS. ADAC is situated at SMHI, Norrköping, and ATESTO in Stockholm. See fig 2, Automatic Data Acquisition System.

2.2.1 ADAT

The ADATs used in the existing networks are of two types. They are built up in either a modular system (ADAT-1) or an integrated one (ADAT-2), both according to the "European Standard System". The modules are mounted in 19" frames with power supply for 220 VAC or 24 VDC.

The modules are interconnected by a data highway. All communication between different modules is controlled by a control unit which includes an INTEL 8080 micro processor. The data highway contains lines for data, addresses and "handshaking" signals. The control unit can for example be programmed for the following functions:

- scanning of sensors
- communication with remote sensors
- calculation of mean values
- optional advanced data reduction
- telegram formatting
- communication with central station (CS)
- local presentation on printer or CRT display

For the communication with ADAC a start/stop character communication according to ISO 1177 is used. The code used is CCITT No 5. The same type of interface and code can be used for connection of remote sensors or local printers/displays.

A typical ADAT-1 consits of the following units:

- power supply AC/DC or DC/DC
- control module including real time clock
- memory PROM/RAM (can be expanded in steps of 4 kbyte to 64 kbyte)
- communication interface MUX/ADC (can be expanded in steps of 16 channels)
- signal conditioner for each sensor(if the sensor does not have standard output 0-(4)-20 mA or 0-5V)
- digital output (if the ADAT also has a control function., e.g., controlling a suntracker

In the integrated system, ADAT-2, the processor, memories, MUX/ADC and communication interfaces are built up at one double size module according to the "European Standard System.

A new integrated, low power system ADAT-3 will be used in the PROMIS-600-system. This terminal has in its basic concept
- 8 analog inputs
- 2 pulse inputs
- 8 digital inputs
- 8 digital output
- 3 serial input/output for communication

In ADAT-3 the microprocessor INTEL 8085 is used.

For more powerful applications, e.g., systems for meteorological measurement, presentation and communication at airports, the microprocessor INTEL 8086 is used. See reference (6).

2.2.2 MICROPROCESSOR SOFTWARE

The program modules are principally written in high level language (PL/M) and designed in a uniform manner to simplify documentation, declarations and program code. When producing new program modules, great importance is placed upon the fact that the programs can generally be used for different categories of use, e.g., that a program for vectorial integration can be used for current meters and wind gauges. For software development SMHI has procured an INTEL NDS-II-multiuser system; see figure 3. The software library consists today of some hundreds of modules, and when the programmer gets the specification for a new automatic station, it is a simple task for him to link the station software from the stored software modules. Amongst all the software modules, modules for the following parameters can be:

Wind - rolling vectorial mean value
- rolling speed mean value
- rolling direction mean value
- extremes of direction and speed
- variance of direction
- energy

Temperature - water/air
- extremes 12h and 24h with time marking
- profiles

Humidity
Dew point
Air pressure - reduction
- trend

Precipitation
Sun radiation - direct
- global
- net
- control software for suntracker

Water current (same as for wind)

Water conductivity
Sea waves - significant wave height
- average wave period
- wave energy by FFT (Fast Fourier Transform) see reference (7).

Water level - rolling mean value
Local presentation - printer/display
Semi automatic telegram editing - SYNOP
- METAR
- METREFORS

Sensorcheck

3. AN EXAMPLE

One application of the Swedish Microprocessor Based Data Aquisition System is the Ocean Wave Energy Program. This application needs both a lot of measurements in a severe environment and data control, reduction and compution in real time. Therefore it can serve as an example of the possibilities when using a microprocessor based system.

The motion of the ocean consists of many different waves with different amplitudes, wavelength and direction interacting in a complex system. To ensure that an observation is correct, an observation period of a minimum of 10 minutes is required in accordance with WMO recommendations.The choice of a good sensor suitable for the specific environmental conditions is important. In Swedish coastal waters the ice conditions are severe every winter. For safe operation during the whole year, it is more or less impossible to use a sensor at the water surface. For special applications in ice free periods an accelerometer bouy is used, but the main sensor is an inverted echo-sounder placed at the sea bottom. The function of the echo-sounder is shown in fig. 4. A transmitter sends a sound signal at a frequency of 1 MHz upwards. This signal is reflected at the surface water/air and detected at the bottom by the sensor an interval later. Knowing the traveltime and the speed of the sound in the water, the water level is easily calculated. By frequently repeated measurments, in this case 10 times every second, one gets a good picture of the water level changes in one spot. During the 17 minute interval, used, 10.200 measurements are taken. At windy situations many air-bubbles are drawn into the water. A lot of different reflexions from these bubbles are detected by the sensor. The software in the ADAT is used to determin the correct reflexion. (see figure 5.)

Another complication for this computer network is that a radio link extension is needed for data transmission from the observation platform to shore. Only a limited number of frequencies are available in the coastal areas of Sweden. For this network SMHI has been allowed to use 5 sec. every hour for data transmission, which is not enough for safe data transmission of all of the 10.200 waterlevels. Therefore, the ADAT:s microprocessor is used for data reduction. The amplitude of a specific frequency can be calculated by Fourier analysis. In other cases, we use the Fast Fourier Transform (FFT) technic to determine the amplitude at 30 fixed frequencies. The data obtained from the wave gauge programme and transmitted to SMHI are as follows:

A. Wave parameters
- 30 spectral densities (2 of these are average values over the 5 + 5 components in the frequency interval 0,52-0,66 Hz where energy seldom appears)
- The 5 highest waves
- Significant wave height, calculated from the 10:th highest wave
- Mean period, calculated from the number of "zero-downcrossings"

B. Check parameters
- Total number of rejected samples in the observed time serie
- Average height of the rejected samples
- Standard deviation of rejected samples
- Number of rejected measuring series
- The average value of the calculated mean water levels
- Standard deviation of mean water level

Together with other measurements, wind speed and direction, air temperature, current speed and direction, etc., and ADAT control parameters, some 50 figures can easily be transmitted in the time interval that is at our disposal. This data reduction from 10.200 figures to 30 is no loss of information since the original data set can always be reconstructed.

4. EXPERIENCES

The Swedish Meteorological and Hydrological Institute started to use microprocessors in 1977. Today approximately 60 fieldstations (ADAT) equipped with microprocessors are in operation and another 40 fieldstations will be taken into operation in the next 2-3 years. Our experiences are, as shown by the example above, that rather complicated data collection projects can be solved in a relatively simple way by using microprocessors and software instead of only hardware. The technique is so powerful that the limitation is rather the lack of creative imagination by man than the capability of the computer. We have also found that it is an enormous advantage to have the maintenance of both the hardware and the software in our own organization. This means that the data acquisition system must be built up in a modular way, from the sensor to data storage.

Development and maintenance of software is rather expensive. Costs can be reduced by using the same type of microprocessor in so many applications as possible and to build up a library of calculation modules. By combining the modules, a suitable program for a new station can be produced with a minimum of costs.
It is preferable to only use one central station to collect data. The need of real-time data in local areas, for instance airfields, pilot houses, road authorities, mountain cable-car stations etc, is satisfied by locally placed displays. Updating is done every time the central station makes a data collection or from the displays if needed.

5. DEVELOPMENT IN THE NEAR FUTURE

During the winter period 1984/85 a microprocessor equipped water quality monitoring station will be developed. As a field station an ADAT-3 will be used. Measurements of pH, water temperature and water level will be used for calculating the need of lime in acid river-water (se fig 6). The ADAT-3 will control the lime dosage from the container and check-up the results down stream. The status of the river, the functioning of the liming equipment and the resulting pH value will also be transmitted in real time to SMHI and displayed at the local environmental authority.

References

1. Hovberg, T. 1980, The Automatic Surface Observation System in Sweden - Experience and Future Outlooks, WMO. Instruments and observing methods, Report No 1, Automated Meteorological Systems, 213-220.

2. Kvick, T. & Zimmerman T. 1980, The Meteorological Measurement System at the Sites of Two Prototype Wind Turbines in Sweden. SMHI, Norrköping, Sweden.

3. Dahlgren, L. & Magnusson, S. 1983, A Solar Radiation Network in Sweden Using Automatic Data Acquisition Techniques, American Meteorological Society: Fifth symposium Meteorological Observations and Instrumentation, Preprints, 65-67.

4. Bodin, S. & Magnusson, S. & Moen, L., PROMIS-600 The Pilot Project for the Future Swedish Weather Service. SMHI, Norrköping, Sweden.

5. Hovberg, T & Udin, I. 1984, Manual Data Entry Device for Synoptical and Climatological Observations in Sweden, paper to be presented at the WMO TELEMO conference on instruments and observation methods in the Hague, September, 1984.

6. Larsson, A. & Hovberg, T. 1983, SAWO-An Efficient New Weather Observation System for Airports Used in Sweden by the Board of Civil Aviation, American Meteorological Society: Fifth Symposium Meteorological Observations and Instrumentation. Preprints, 421-426.

7. Mattisson, I. 1978, A Wave Measuring Project for Wave Energy Prospecting Purpose, COST 43 Seminar December 4-6 1978 Lisbon.

8. WMO, Handbook on Wave Analysis and Forecasting No 446 WMO Geneva Switzerland.

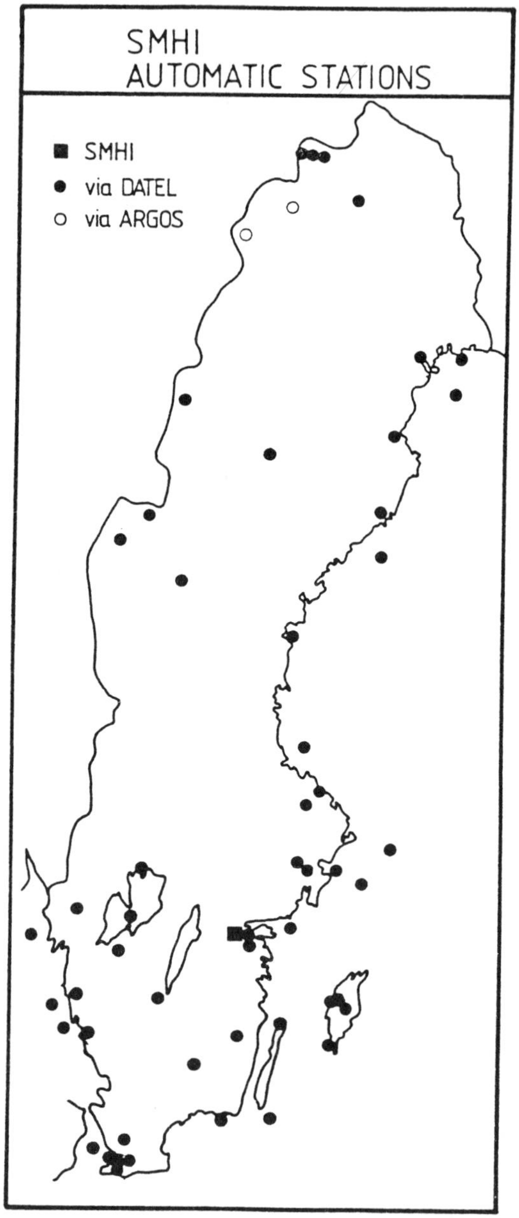

Fig 1. SMHI. Automatic station network

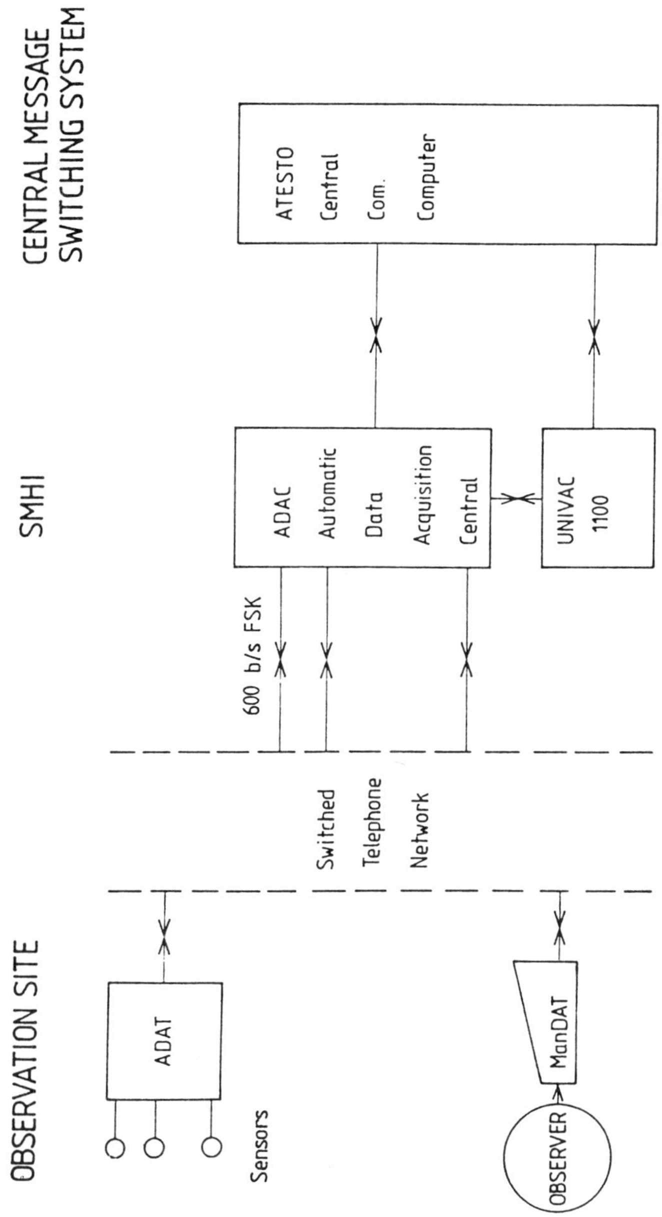

Fig 2. Automatic Data Acquisition System, ADAS

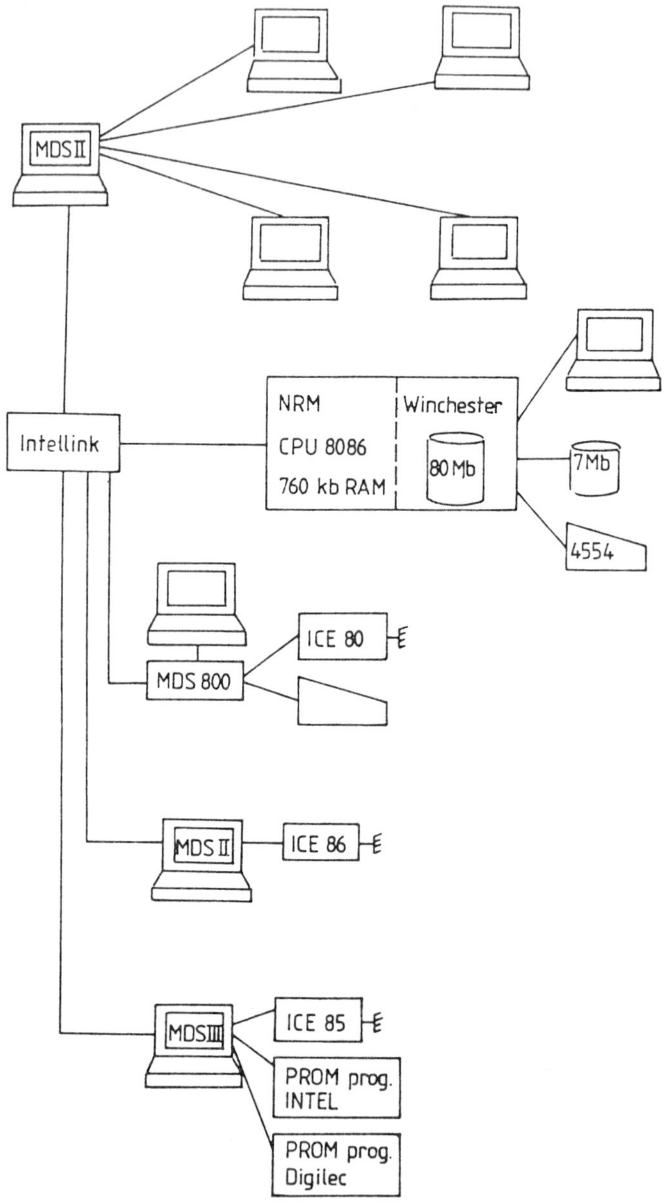

Fig 3. SMHIs system for micro processor software development, INTEL NDS-II

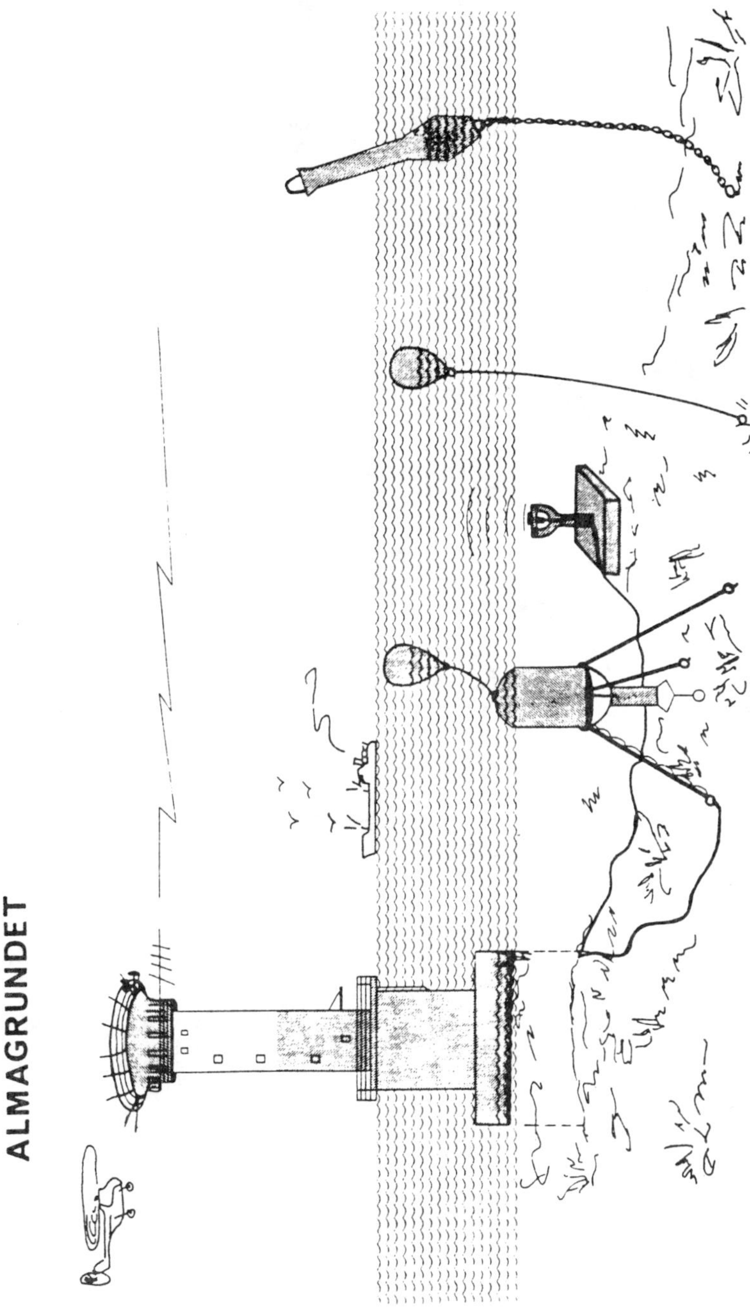

Fig 4. The inverted echo-sounder is placed on the sea bed at a depth of 25 m and about 100 m from the lighthouse

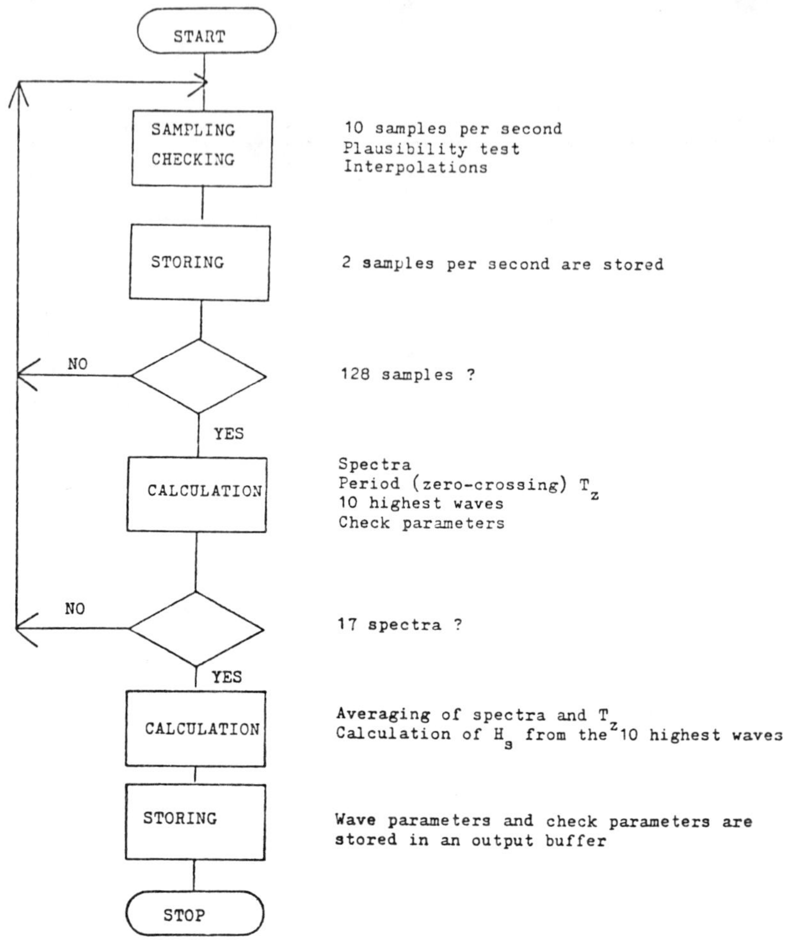

Fig 5. Generalized flow chart of the microprossor program

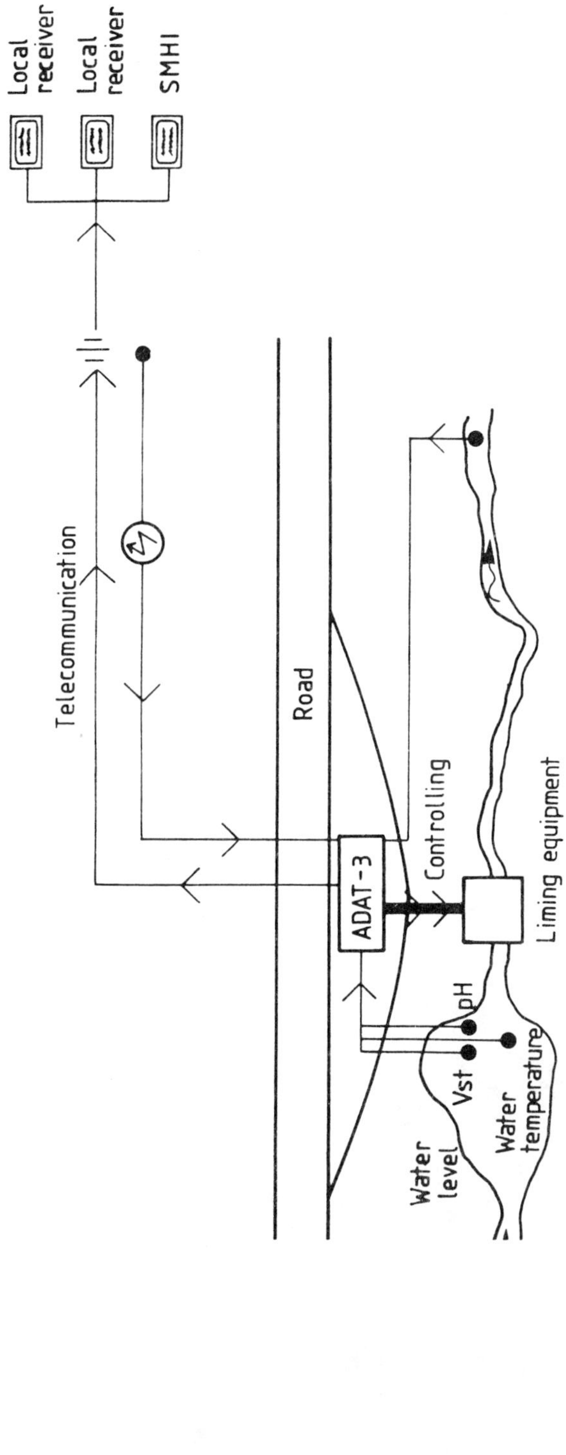

Fig 6. Controlling of liming equipment and transmission of information

SHORT TERM SNOWMELT FLOW FORECAST SYSTEM

Humberto Peña T.*
Brahim Nazarala G.*

SUMMARY

The river regime in the central part of Chile is greatly affected by the accumulation of snow and ice in the Andes Cordillera during the winter months (April-September) and its subsequent melting during the spring-summer period.

The heavier-than-normal precipitation of 1982 made it advisable during the months of snowmelt to set up a river flow telemetering and flood forecasting system. For this purpose, data collection platforms based on the GOES satellite were installed at river gauging stations. Because weather forecasts are not available more than 24 hours in advance, the forecast gave the maximum discharge to be expected for the three following days assuming the most favourable conditions for snowmelt. Thus the range of possible variations in future flows was obtained.

Flows were obtained through the use of a very simplified hydrological simulation model. This model subdivided the catchment area into areas of equal elevation and considered as input data the flow of the day prior to the forecast, maximum daily temperatures at a point in the catchment area and the snow line. In addition, an adjustment algorithm was incorporated into the model to take account of the previous days' results.

This particularly simple forecasting procedure was successfully used during the 1982/83 snowmelt season.

The simulation model was developed in APL language and programmed into a 64 K microcomputer.

1. BACKGROUND

The central part of Chile has a mediterranean climate, with a short, rainy winter and a long dry season which lasts roughly seven months. Its main water resources derive from the Andes Cordillera, which guarantees the water supply to agriculture in the dry spring and summer season.

In 1982, the largest accumulation of snow in the last thirty years was recorded. This caused great concern about the possible size of snowmelt flows and prompted the development of the forecast system described here.

The precise aim of the system was to provide frequent periodical information about flows recorded on rivers and flow forecasts for the following days. In this instance, because the flow is directly affected by weather conditions that make the snowmelt possible, and because weather forecasts are available only 24 hours in advance, it was only possible to make estimates assuming weather conditions particularly favourable for snowmelt on the following three days. In this way, a limit could be placed on expectations for the immediate future.

*Sub-department of Hydrological Studies. General Directorate of Water Resources, Chile.

World Meteorological Organization (ed.), Microprocessors in Operational Hydrology, 155-161.

2. CONSTRUCTION OF THE MODEL

In designing the model, a condition was imposed that it should be easy to calibrate and that it would use few input data. Furthermore, it was preferable for it not to require much of computation. The model was to forecast the maximum daily flow.

Conceptually, the model can be divided into two phases, one dealing with the melting of the snow cover and the other with the resultant runoff in the river.

2.1 Snowmelt model

As regards snowmelt, the model was built on the following bases:

(i) Subdivision of the catchment into sections of equal altitude; these sections form the unit of calculation for the snowmelt process.

(ii) The snowmelt for each section is determined daily on the basis of an equation of the type

$$D_i = a\,(TX_i - T_b) \text{ for } TX_i \quad T_b \qquad (1)$$

where:

D_i = Daily snowmelt for section i
TX_i = Daily maximum temperature in section i
a = Degree-day snowmelt factor (parameter)
T_b = Reference temperature (parameter)

The effect of spring and summer precipitation on the snow cover was not taken into account.

(iii) Maximum daily temperatures have a single gradient with respect to altitude constant throughout the entire period.

(iv) The area within the catchment that actually contributes to the snowmelt is bounded at the lower level by the snow line, which is an input datum for the model. The upper limit is set by the following condition that has to be met for snowmelt to occur:

$$SDi \quad C\,(Ah_i - B) \qquad (2)$$

where:

SDi = Accumulated value of the snowmelt obtained from equation (1) for section i, starting from the first D_i 0

C = Factor proportional to the snow accumulation for the year

h_i = Altitude of section i

A,B = Parameters

Equation (2) makes it possible to consider the thermal storage in the snow cover during the winter months and the retention of water liquid both in the snow and in the soil.

2.2 Runoff model

On the basis of an analysis of a typical snowmelt hydrograph, and considering recession curves of the Qe^{-kt} type, equations (3), (4) and (5) were deduced for the maximum daily flow in accordance with the outline in Figure 1.

$$QP\ (n) = \frac{2k}{2 + k\ (1 - tm)}\ QF\ (n) + e^{-k}\ QP\ (n-1) \qquad (3)$$

$$QP\ (n) = \frac{2k}{2 + k\ (1 - tm)}\ QF\ (n) + e^{-k\ (1-tm)}\ Qm\ (n) \qquad (4)$$

$$Qm\ (n) = QP\ (n-1)\ e^{-ktm} \qquad (5)$$

where:

QP (n), QP (n-1) = Maximum flows for days n and n-1 (m^3/s)
QF (n) = Volume of snowmelt for day n (m^3/s day).
Qm (n) = Minimum flow for day n (m^3/s)
k = Parameter for the recession equation (day^{-1})
tm = Period of time between maximum and minimum discharge (day).

As can be seen, the coefficients of these equations are determined when parameters k and tm are known.

2.3 Adjustment factor

As the model is to be used during the entire snowmelt period, it is possible to use the information from previous days in the forecast in order to improve the calibration of the equations. Therefore, an algorithm was introduced whereby, by comparing the moving averages of measured and simulated QF values in the last few days, the value of the degree-day factor obtained in the original calibration can be adjusted.

3. COMPUTERIZED ASPECTS

The theoretical model described above was run on an IBM-5110 microcomputer with the following configuration: a 64 K main memory using BASIC and APL languages, a slow 132 characters per line printer and a dual diskette unit using 1.2 MB floppy discs.

The systems controlling the running of users' programs both in BASIC and in APL, take up between 5 and 6 K of the main memory, leaving the rest of the memory free for the development of applications.

The hydrological model was developed in APL language with a main function called "FUSION", that runs checks on data and monitors the performance of the other functions. The system has components to update variables with input data (FUSACT), to calculate the melt (FUSMOD) to calculate the peak flows on the basis of the melt (FUSPEAK), to make forecasts covering "n" days, or to make correlations between measured and forecast flows (FUSPRON). The system's operation is shown schematically in Fig. 2.

The information, both input and output, is handled through variables, each one of which has a set or series of data of the same type. Most of these are one-dimensional arrays with daily data of a given element, such as: daily temperature at the base weather station, the height of the snow line, the simulated snowmelt flow, the simulated peak flow, etc.

The system may be run for three basic operations: data update, simulation and forecast. Through the first of these, input variables can be updated and the consistency between the length of the various arrays checked. The simulation mode operation calculates the snow melt and peak flows, and forecasts covering 1, 2 ... n days starting from each day with information. If desired, correlations can be made between observed and measured flows. The forecast mode operation draws up a forecast covering "n" days based on given values of the height of the snow line, the weather station temperature and the degree-day factor, starting from the last day of the data series.

The six functions in the system take up 13116 bytes in the IBM-5110 microcomputer. The functions that calculate correlations between observed and simulated flows require an additional 7760 bytes.

Depending on the volume of input and output data, the memory capacity of this microcomputer may be exceeded. Should this happen, the problem can be overcome either by eliminating data as these are used in the system or by carrying out processes in stages.

In this microcomputer, for 15 sections, the processing time is roughly three minutes per month of statistics.

The system does not carry out direct operations on peripheral recording equipment and only uses the basic APL functions. There are therefore no particular problems about transferring the model to another computer.

4. USE OF THE FORECAST SYSTEM

During the 1982/83 snowmelt, the forecast system was used on the basin of the rivers Maipo and Aconcagua which, in the reaches running through the Cordillera, have a catchment of 4950 km^2 and 2400 km^2, respectively. Both river flow through the most densely populated part of the country, including the two largest cities (Santiago and Valparaíso).

4.1 Collection of field information

The various data required to operate the model were obtained through diverse means, such as direct observation, conventional weather stations with radio transmission and automatic stations. It is worth describing in greater detail the system used with data collecting platforms (DCPs), to record levels in the rivers Aconcagua and Maipo.

The DCPs installed were made by LaBarge, Inc. with 10 W transmission using a crossed dipole antenna and were connected to a float-type recording gauge. The low energy consumption made it possible to operate this system for a period of five months using two 80 AH batteries with no need for special recharging.

Levels were recorded every half hour and the information, stored in the memory of the DCP, was sent every three hours to the GOES-E satellite and picked up at the Peldehue Satellite Tracking Station. The signal was automatically demodulated and fed into a microcomputer which converted the river level into discharge. In addition, the information was translated into Bandot code for transmission over a microwave link system to the receiver antenna at the Santiago city centre. From there, by wire, data was finally printed out on a teletype in the emergency co-ordination centre. The transmission of information was handled by the Space Studies Centre of the University of Chile.

It should be noted that there were no particular problems encountered in the transmission of the data via the DCP during the season mentioned, and that, on the contrary, timely and reliable information was obtained.

4.2 Calibration and use of the model

The hydrological model was calibrated with satisfactory results by taking into account the information obtained during years with a large accumulation of snow and other past data available about the hydrology of snow in the area.

Prior calibration made it possible to use this system during the 1982/83 snowmelt, by taking into account the following input data: flow on the day of forecast, snowline, and temperature at a station in the catchment area. As pointed out under Background above, given the lack of temperature forecasts in the area, for the three following days the most adverse temperature conditions, drawn from historical records,were used. The graphs in Figures 3 and 4 show a comparison of maximum daily flows measured and simulated in the following cases:

(a) A three-day forecast made assuming the temperatures of the following days are known. This graph reflects the error due to the simulation model.

(b) A three-day forecast as above, but taking into account the most adverse temperatures drawn from historical records. This graph reflects the error due to the model plus the difference caused by forecasting with maximum temperatures higher than those recorded. The forecasted discharges should be the envelope of measured discharges during the forecast period.

5. CONCLUSIONS

The experience obtained with this study leads to the conclusion that, for short-range forecasts of snowmelt flows simulation models can be used which are simple enough to run on a microcomputer, yet which produce perfectly acceptable results for practical purposes. In addition, automatic information collection systems proved to be efficient and of great operational advantage in flow forecasting.

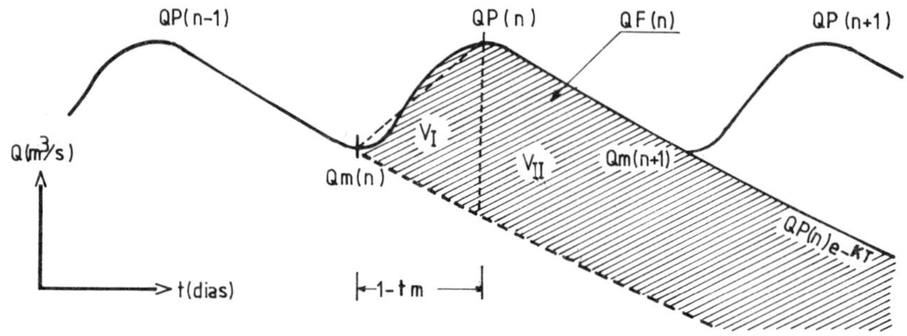

Figure 1. Snowmelt hydrograph.

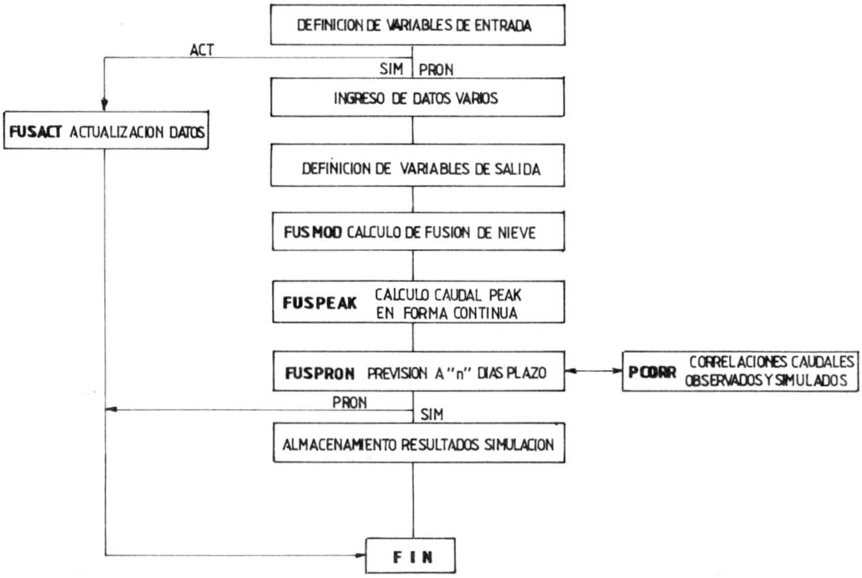

Figure 2. Computation scheme of model.

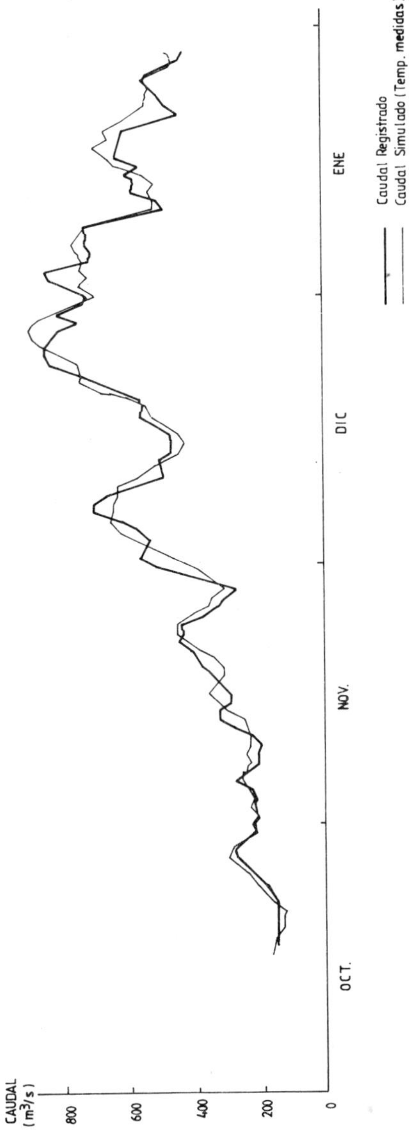

Figure 3. River Maipo at El Manzano; 3-day forecast.

3-day forecast

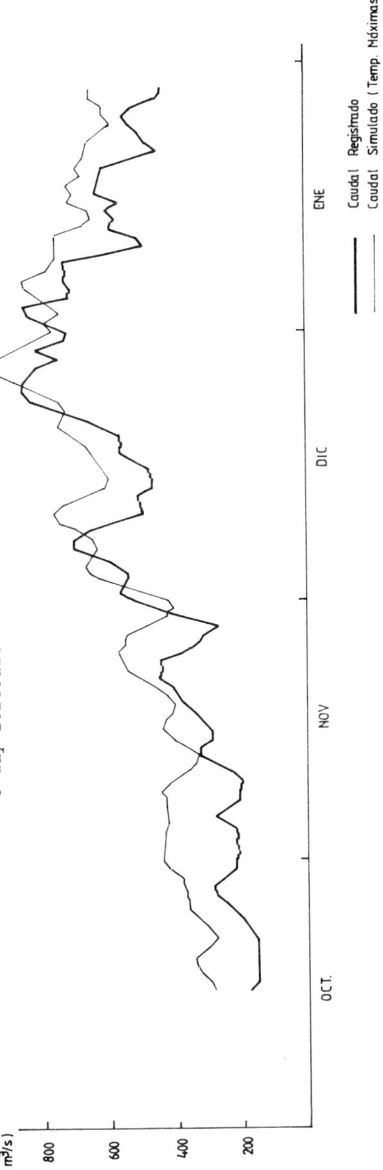

Figure 4.

OPERATIONAL FLOOD FORECASTING MODELS FOR MICROCOMPUTERS

P. Bartha*, A. Szöllősi-Nagy** and K. Harkányi***

INTRODUCTION

Operational hydrological forecasts are being made in Hungary for several decades by the National Hydrological Forecasting Service (NHFS) residing at VITUKI (Water Resources Research Centre) and by the Regional Water Authorities (RWA) of Hungary. The methods used were demanding of tiring and cumbersome manual work from expert hydrologists. These methods, in spite of the sophisticated use made of the information of the whole upper river reach, are based on simplified multivariate linear regression techniques. Because of the lack of computers, they had been given in the form of co-axial graphs by Szesztay (1957). These graphs have been used for more than twenty years with a rather good efficiency but without a possibility to follow both sudden and slow changes in the water regime.

With the increase of the national assets an increasing attention has been paid towards the development/improvement of forecasting technologies. In 1979 a nationwide project was launched to promote developments in hydrological forecasting. One of the main objectives of this project was the development of a real-time computerized forecasting system for the Danube River. (More on the project can be found in Bartha and Szöllősi-Nagy; 1982).

At the beginning of this development project it was set that an operational forecast method should preferably satisfy the following requirements:

(1) It should consider the physics of the forecast process as much as feasible;
(2) It should explicitly take into account the type, size, and propagation of uncertainties involved in any forecasts;
(3) It should be adaptable either to natural or man-made changes in the environment;
(4) It should yield optimal forecasts according to preset performance measures;

---

* Chief, National Hydrological Forecasting Service (NHFS), Water Resources Research Center (VITUKI)

** Head, Department of River Hydrology, VITUKI

*** Research Associate, NHFS, VITUKI

World Meteorological Organization (ed.), Microprocessors in Operational Hydrology, 163-178.

(5) It should be suitable for incorporating additional information without changing the structure of the forecast method;
(6) It should be suitable for filling in missing information;
(7) It should satisfy the criteria of numerical stability, parameter sensitivity, and solution convergence;
(8) It should be easily implemented on relatively modest computers, usually available for operational forecast purposes;and
(9) It should be connected to and imbedded into the operational models of water resources systems without significant modifications.

By studying the existing forecast methods under different conditions of application, it becomes clear from the comparison of operational forecast requirements with the ongoing forecast methods, that a universal operational forecast method, i.e.,the one that is "best under all possible conditions" does not exist.Likely, it will neither exist in the near future. The emphasis must be placed on the term "operational forecast" since current data bases and theories do not permit all the processes of the hydrologic cycle to be included in a unified, general forecast method. However, the operational forecast methods should be flexible in the sense that models should easily be tailored to particular conditions, whether dealing with small catchments or large river basins. It was therefore a plausible approach to develop forecast approaches which are of the modular structure. Modularity is understood in two ways, both as horizontal and as vertical modules. Horizontal modules are to be set up according to geographic and hypsographic characteristics of a watershed. The aim is to circumvent problems associated with very detailed spatial (or areal) variation of variables. The vertical modules describe the hydrologic processes within each horizontal module of a catchment area. As a principle: Each module obviously should somehow be physically based. This does not mean, however, that unavoidable problems of uncertainty should be neglected. On the contrary, an operational forecast method should always include the consideration of inherent uncertainties associated with any forecast. These uncertainties originate from:

(a) limitations in selecting the numbers and types of modules;
(b) imperfect model structure;
(c) observational errors;
(d) approximation of distributed variables by lumped variables;
(e) inefficiency and bias in parameter estimations;
(f) sampling variations, etc.

Consequently, the modular system is a joint composition of physical structure and stochastic scheme, in which the physical structural part takes care as much as feasible of hydrodynamics of the system while the stochastic part describes the character of residuals of the structural model.

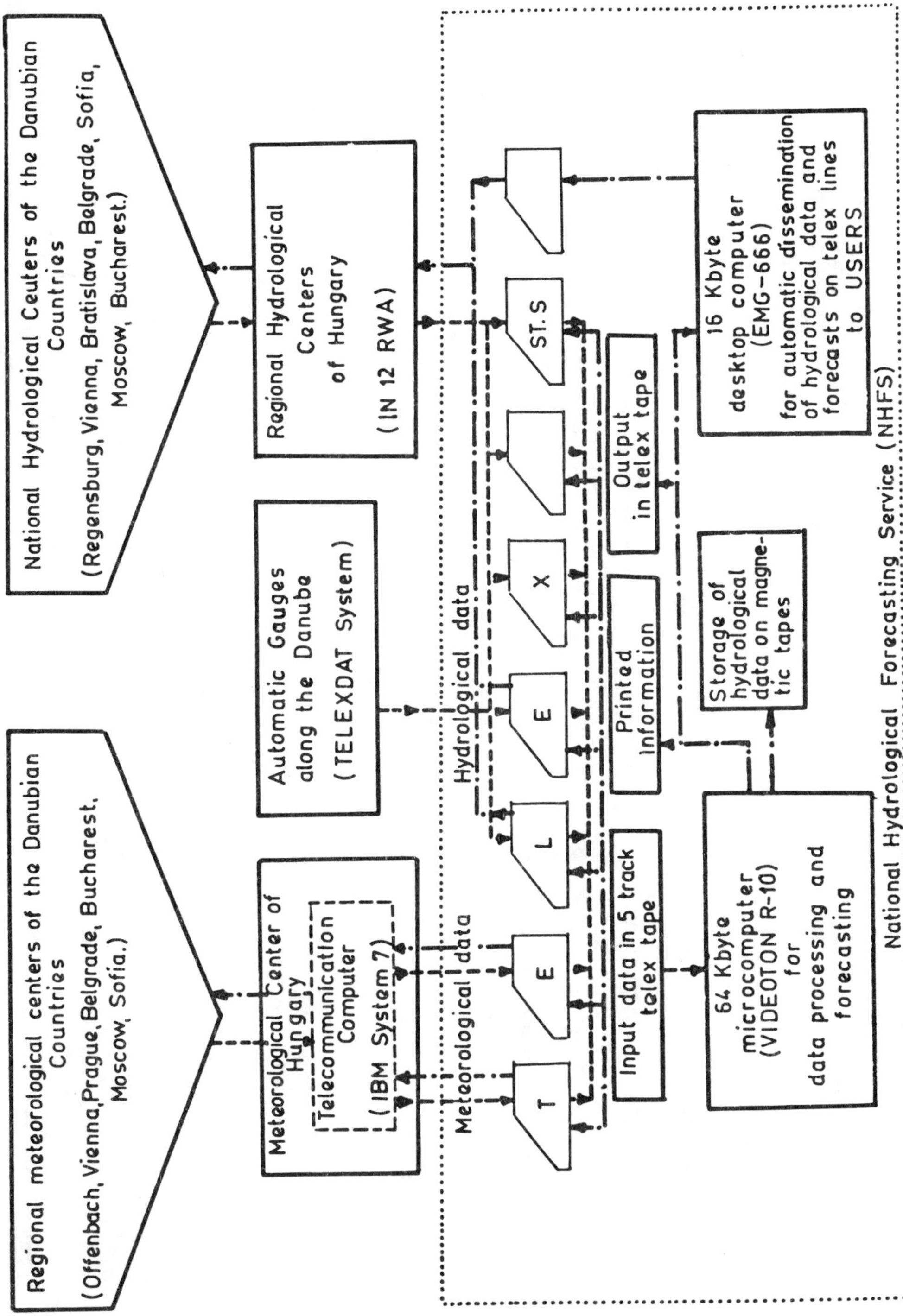

Fig. 1. Information flow diagram of the Hungarian NHFS

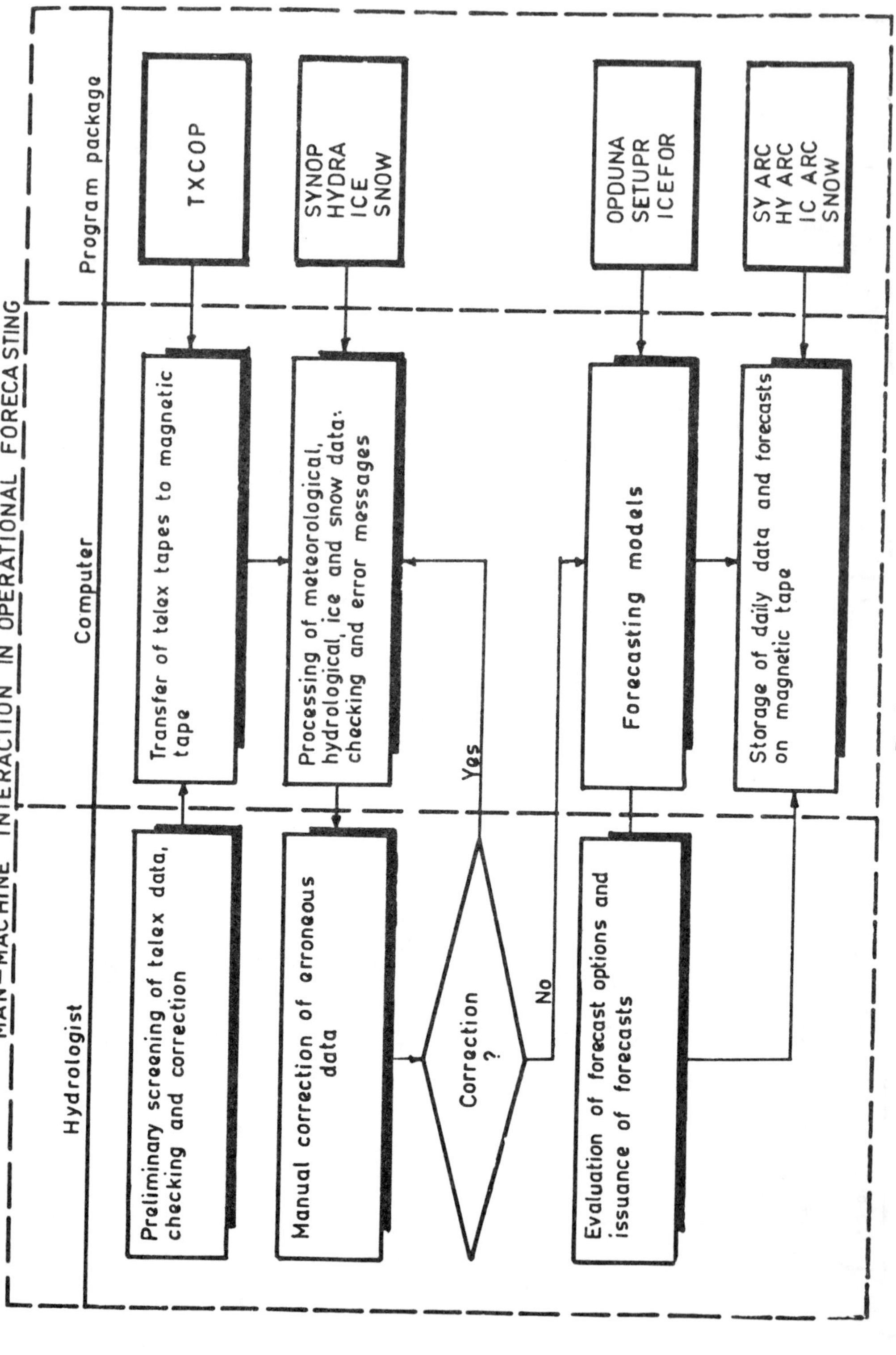
MAN-MACHINE INTERACTION IN OPERATIONAL FORECASTING
Hydrologist
Computer
Program package
Preliminary screening of telex data, checking and correction
Transfer of telex tapes to magnetic tape
TXCOP
Manual correction of erroneous data
Processing of meteorological, hydrological, ice and snow data: checking and error messages
SYNOP HYDRA ICE SNOW
Correction ?
Yes
No
Evaluation of forecast options and issuance of forecasts
Forecasting models
OPDUNA SETUPR ICEFOR
Storage of daily data and forecasts on magnetic tape
SY ARC HY ARC IC ARC SNOW

Fig. 2.

Even in the case if one posses a structural-stochastic forecasting model an optional back-up model is needed for reasons of safe operation.

The NHFS receives around ten thousand data daily (3.65 million data annually). In order to process this amount of information effectively the use of computers is needed. When setting up a computerized forecasting system at NHFS the following requirements were considered:

- All the processing is to be based upon a 64 K byte microcomputer;
- Processing is to be organized in such way as to allow for human intervention beside the algorithmic checking and correction of incoming data;
- All the processed data are to be stored in archives; and
- The processing and forecasting system should yield forecasts by 08:00 GMT hour.

Figure 1 summarizes the information flow diagram of the NHFS giving indications for the hardware tools used.

The logistics of processing/forecasting is sequential that always permits the intervention of the forecaster. This man-machine interaction, shown in Fig. 2, makes the use of the system flexible.The present authors feel that the dream of a fully automatic/computerized hydrological forecasting system is to be given up and there always should be some room left for the judgement of the operational hydrologist.

It is to be mentioned here, that within the framework of HOMS VITUKI has elaborated a flood forecasting sequence for the HOMS Asia and the South-West Pacific WMO/UNDP regional project (RAS/79/113) on the basis of a General Agreement between WMO and VITUKI. The sequence has been developed for a TANDY TRS-80 Model-II microcomputer with TRSDOS operating system and is essentially the same as the forecasting procedures used by the NHFS using a microcomputer with the standard CP/M operating system (Version 2.2).

This paper puts the emphasis on the organization of the program package,though a short overview of the methods used is also given. Details of the models can be found in Szöllősi-Nagy et al. (1983).

OPERATIONAL FLOOD FORECASTING MODELS USED BY NHFS

The connection of the different flood forecasting components is shown in Fig. 3. Most of the components are directly connected to the Hydrological Data Base (HDB), see Futaki and Vargay (1984, paper presented at this Conference), while some use their own data files.

Gauge relations for flood peak forecasting

The method of corresponding stages is one the simplest methods for short-term operational forecasts. Essentially the method amounts to finding the correlation between stages at the upstream station of a reach and stages of what will consequently occur (after a delay caused by the travel time) at the downstream station. The difference in time of the occurrence of corresponding stages at the upper and lower stations determines how far in advance it is possible to forecast.This depends on the distance between the stations, the morphometric features of the reach and magnitude of discharge. The method of corresponding stages gives good results for forecasting peak discharges and stages.

Gauge relations can be applied most successfully on major streams where the stages rise and fall at slow rates. The results are most accurate if the local inflow between the two cross-sections is small relative to the discharge entering through the upstream cross-section. This is the case over reaches where the river flows in a uniform, regular channel are without major tributaries and appreciable infiltration. The gauge relations involving stage data are of higher accuracy in relatively wide beds where even major changes in discharge entail only small changes in stage.

The model is considered to be satisfactory for water stage forecasting if correlation coefficients between independent variables are low and one of the correlation coefficients (preferably for the variable representing peak stage at the upper station) has a higher value (0.8-0.9) and the multiple correlation coefficient is at least 0.95. The model is considered to be satisfactory for travel time forecasting, if the multiple correlation coefficient is at least 0.65. For the stability of the parameters for gauge relations with only one independent variable a data set with 15-20 flood events as a minimum should be used and the number of flood events should be increased by 7-10 floods for each additional auxiliary variable. In other case the standard mean error of forecast exceeds the standard error of estimate.

One should note that the above described limitations were not devided from theoretical considerations, but were gained from long years of experience. Therefore this is to be considered as a rule of thumb and not a rigid prescription.

Although practice showed that gauge relations can be used mostly for forecasting flood peaks some experience is gained using this technique for continuous water stage forecasts, however, it is rarely used in the practice of hydrological forecasting. When correct discharge data are available more efficient hydrological routing models are used instead of simple regressional techniques.

Coupled structural-stochastic model for real-time forecasting

The developed models for on-line (real-time) operational hydrologic forecast so far are almost exclusively based on the principles of state space analysis (Wood and Szöllősi-Nagy, 1980; O'Connel and Clarke, 1981). The reason for the application of state-space techniques in hydrologic forecast lies in the fact that updating the information on parameter and state variables, required in on-line forecast, can be done easily within this framework. Provided an appropriate state-space model is available, the advanced filtering techniques, such as Kalman filtering, can be used for updating of parameter and state variables by using micro-computers. Continuous forecasting models are usually based upon some simplifications of the Saint-Venant equations describing the gradually varying unsteady flow in open channels. A first order approximation of the Saint-Venant equations was given by Kalinin and Milyukov (1958) who introduced the notion of "characteristic reach".The "characteristic reach" technique replaces the equation of moments by a linear relationship between channel storage and outflow from the "characteristic reach". The solution is formally identical with that of the Nash-cascade (1957) used for rainfall/ runoff modeling. Both continuous models are linear lumped time--invariant representations.

However, the continuous model cannot be directly applied in practice since

- measurement data are obtained at discrete time epochs and,
- data processing is done on digital computers, therefore, a discrete model is required indeed.
  However, the discretization of a continuous model is not straightforward (Klemes, 1978). A discretization scheme should be adequate in the sense that it should
- be discretely coincident (with the continuous model),
- preserve the continuity and transitivity.

The model used in the operational work is the Discrete Linear Cascade Model (DLCM) (Szöllősi-Nagy, 1982) connected to an ARMA type stochastic residual model. The deterministic DLCM and the stochastic ARMA model are put up in a coupled model. Forecasts are updated by using Kalman filtering. A typical example of the model performance is shown in Fig. 4.

Entirely stochastic adaptive forecasting model

As discussed in the Introduction there always is a need for a back-up forecasting model under operational circumstances. This back-up model at NHFS is an entirely stochastic time series model based upon the adaptive self-tuning principle.

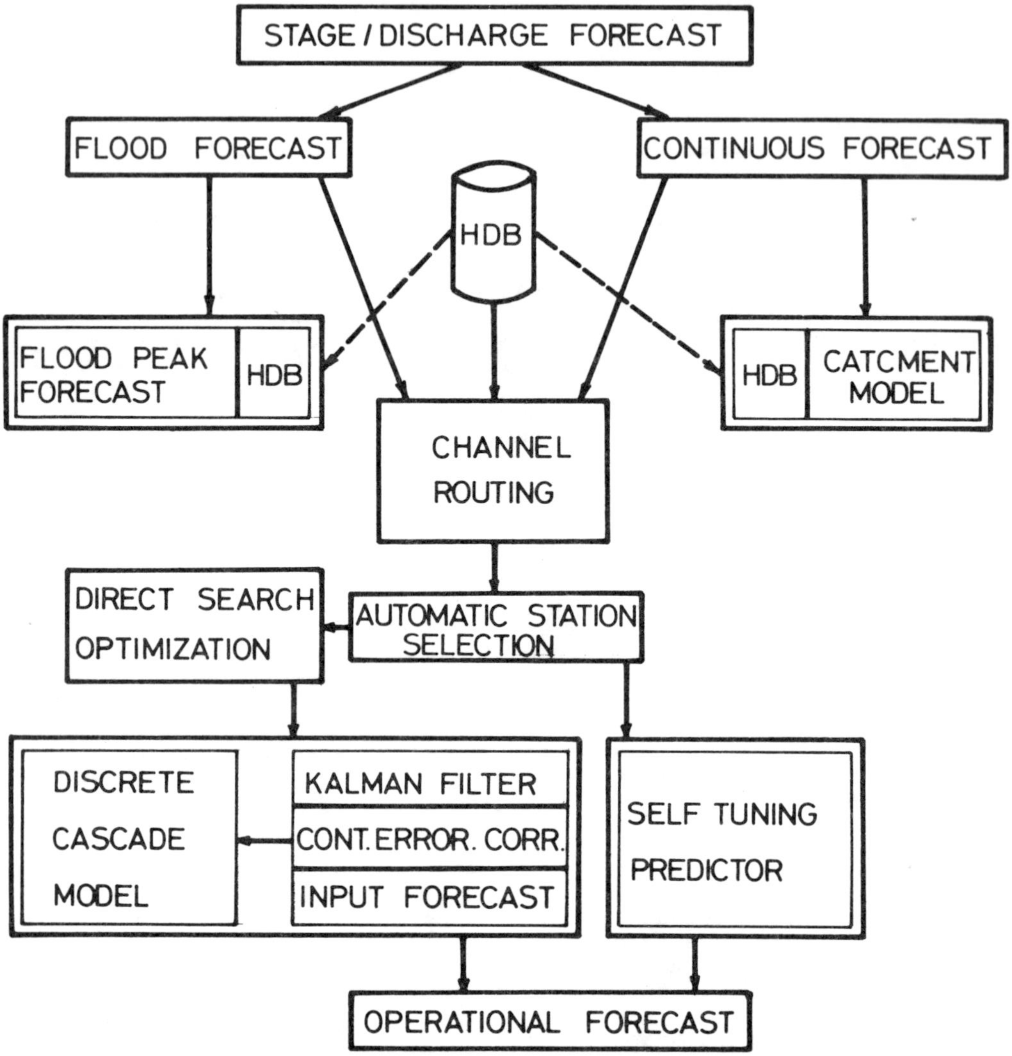

Fig.3.Connection of components in the flood forecasting

The self-tuning predictor model is developed from a simple ARMA time series model. Hydrological applications are already known (Ganendra, 1976; Ambrus and Szöllősi-Nagy, 1981).

The basis of the prediction is the previously done prediction error which is fed back into the model as a variable. Further input variables are the predictions for the given station and measurement data of one (or more) upstream stations(s) as auxiliary variable(s). After receiving the latest measurement data, the parameter vector and a weighting matrix, which is at the same time the covariance matrix of the parameter estimation error, is updated recursively at every discrete time instant and then predictions can be made.

FORECASTING PACKAGE FOR MICROCOMPUTERS

The connection of the models/programs is shown in Fig. 3. All the programs are written in ANSI FORTRAN for microcomputers with CP/M(2.2) and TRSDOS operating system.

Flood forecast

The program package consists of two programs. The first program, FLOODDAT, handles the flood events files. The second program FLOODFOR, determines the gauge relations and enables the operational issuance of flood peak forecasts.

Both programs are set up in an interactive mode. The flowchart of the package is shown in Fig. 5. As the selection of flood data and their checking require the active participation of an experienced hydrologist the programs do not utilize the general purpose HDB but use their own data files the content of which are determined/handled by the forecaster.

Real-time river flow/stage forecast

Programs were written for the coupled structural-stochastic forecasting model.

The operational use of the model requires two stages to perform. The first one is the stage of model calibration, including the coding of the river network parameter estimation etc. The flow chart of this stage is shown in Fig. 6. After the calibration the actual operation of the model starts according to the flow chart presented in Fig. 7.

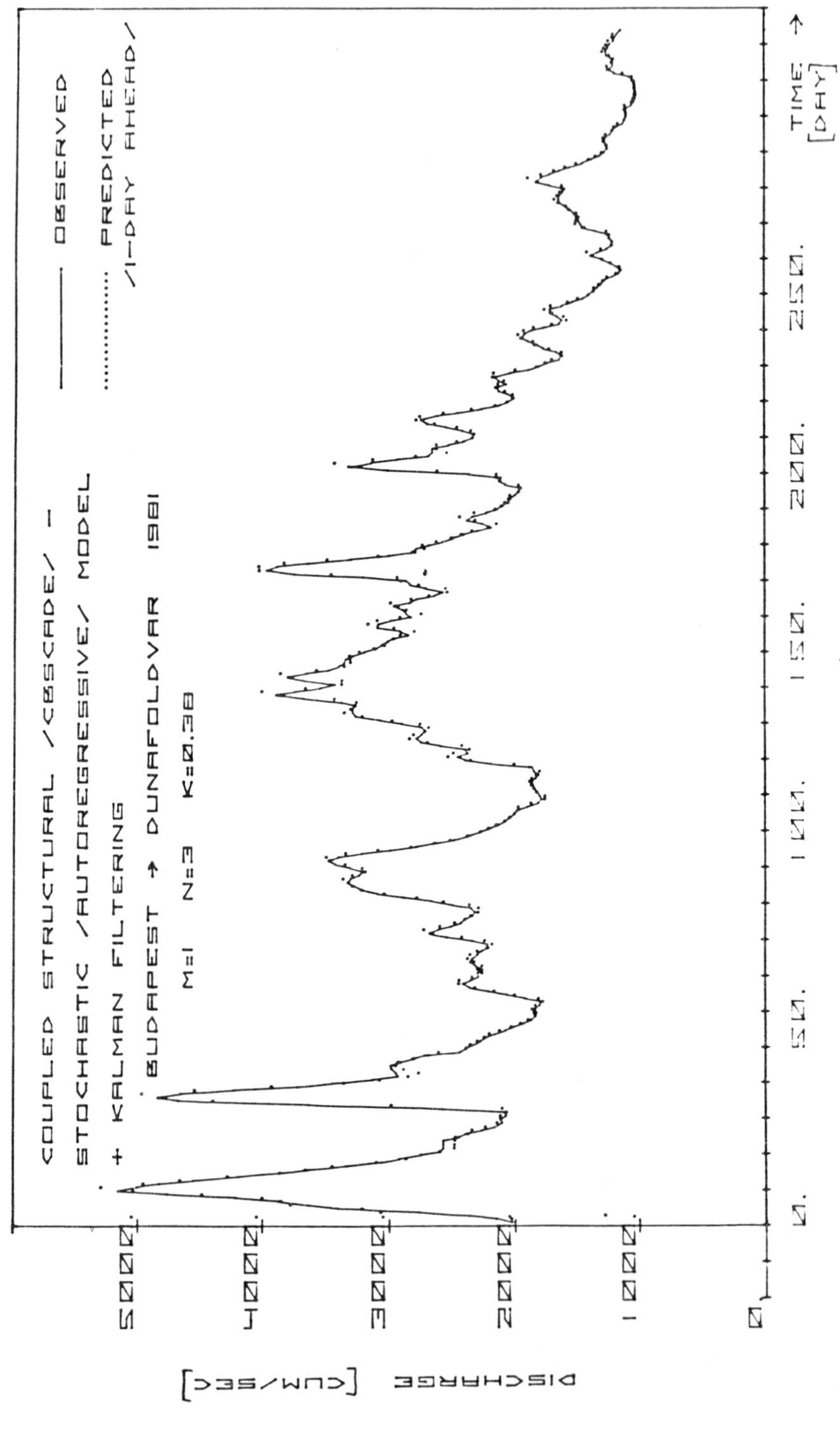
COUPLED STRUCTURAL /CASCADE/ -
STOCHASTIC /AUTOREGRESSIVE/ MODEL
+ KALMAN FILTERING
BUDAPEST → DUNAFOLDVAR 1981
M=1 N=3 K=0.38
OBSERVED
PREDICTED
/1-DAY AHEAD/
DISCHARGE [CUM/SEC]
0
1000
2000
3000
4000
5000
0.
50.
100.
150.
200.
250.
TIME → [DAY]

FIG. 4.

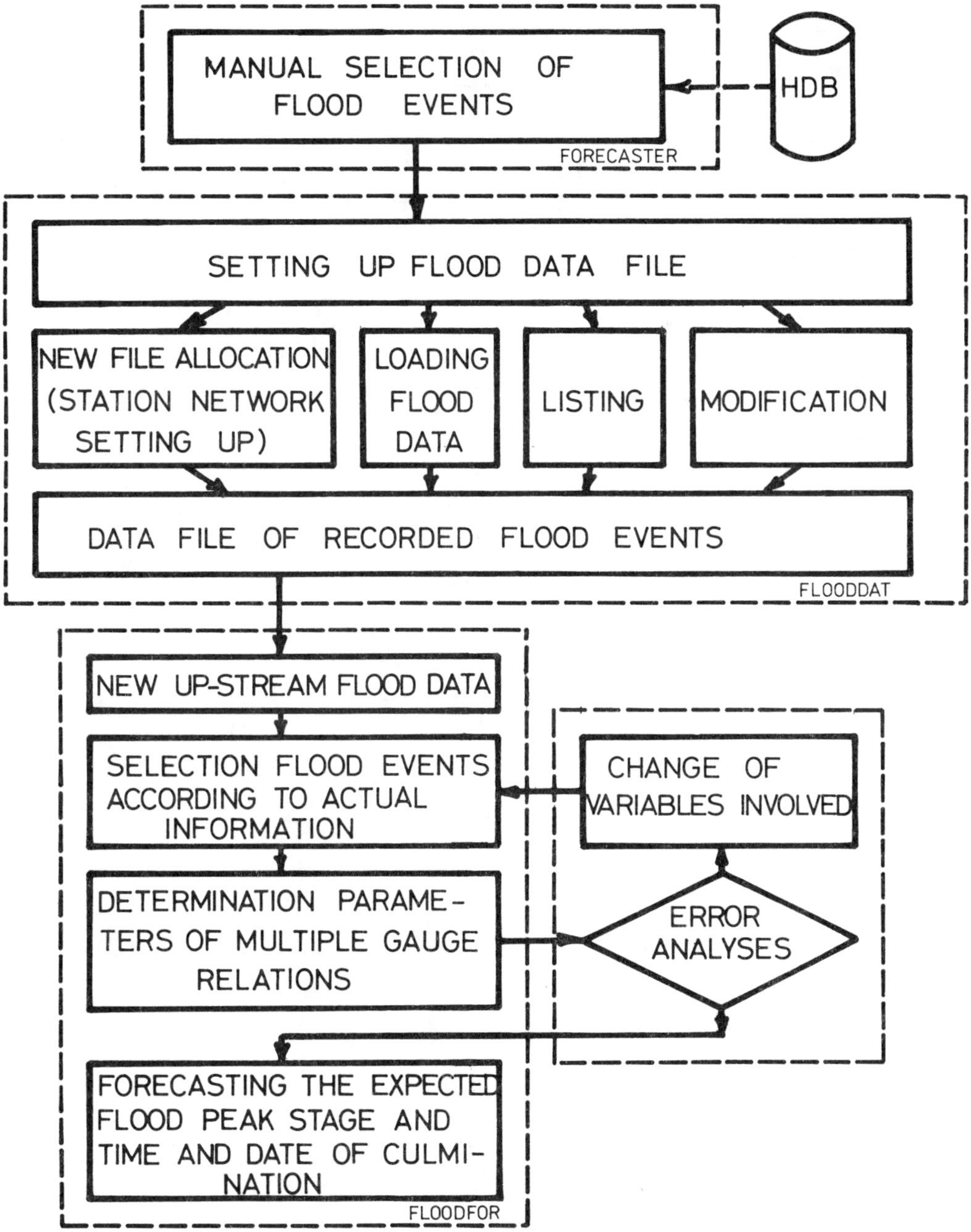

Fig. 5. Flow chart of the flood peak forecasting

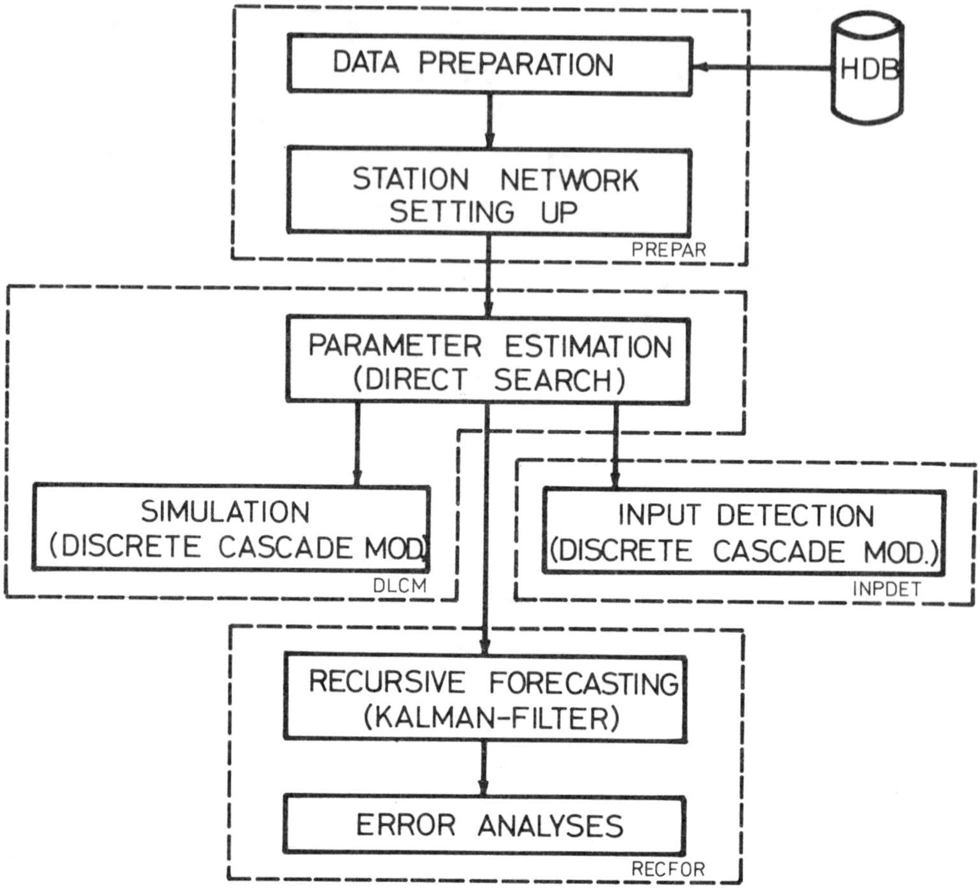

Fig.6. Model calibration and recursive river flow forecasting

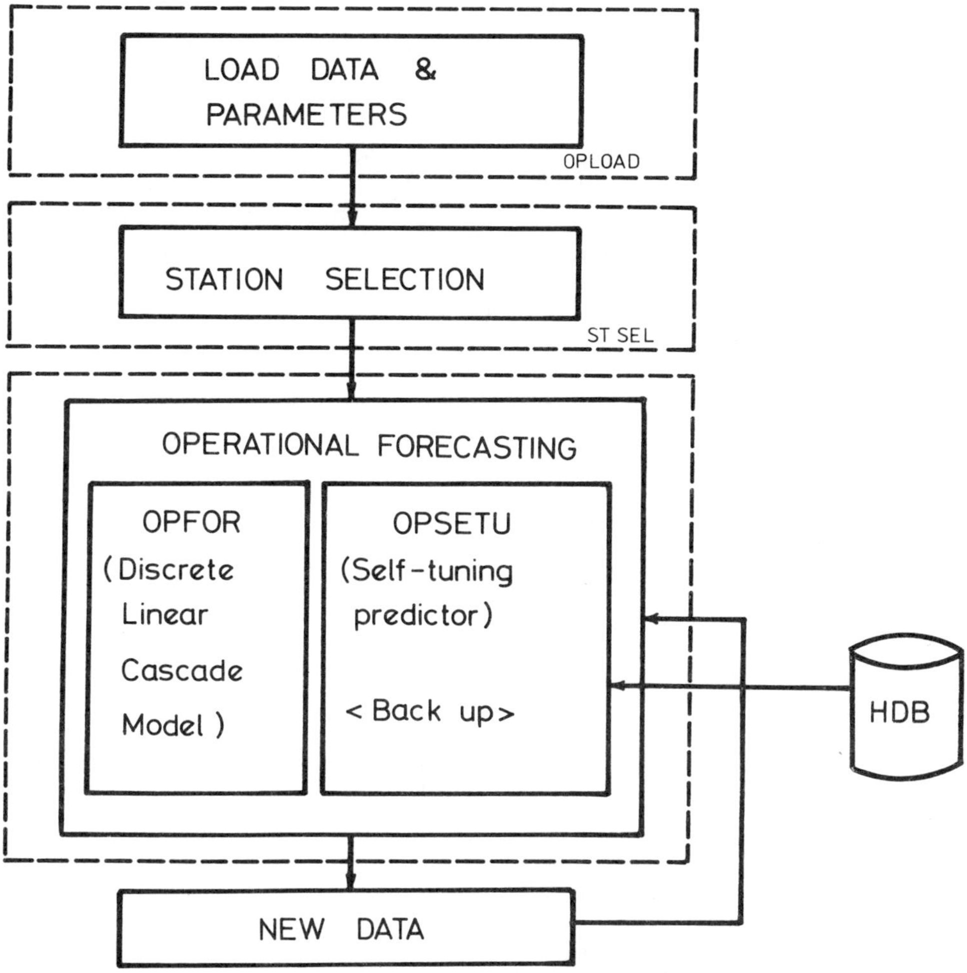

Fig.7. Flow chart of the operational real-time river flow forecasting

The package consists of six programs, viz.

PREPAR	sets up the station network system and prepares data from hydrological data base (HDB) for the programs DLCM INPDET RECFOR.
DLCM	performs parameter estimation, output simulations,and the continuous error adjustment.
RECFOR	performs recursive forecasting using Kalman-filtering and evaluates forecast error.
INPDET	performs the task of input detection.
OPLOAD	loads data for OPFOR program.
OPFOR	performs operational forecasting.

Self-tuning predictor

The self-tuning predictor is used as a back-up model. The package contains several programs.

The first program, ST, is a self-containing one where the user should create data files. The other two programs, PRESET and OPSETU, are connected to the HDB. The way these programs are connected to the operational forecasting system is shown in Figs. 3 and 7.

DISSEMINATION OF FORECASTS

The last element of forecasting activities is the dissemination of forecasts among the users. In forecasting stages for River Danube beyond the interests of water resources management, including flood protection activities, the particular objectives of international shipping are to be considered as River Danube is an international waterway. Due to this, several forms of disseminating hydrological forecasts have been developed. The hydrometeorological centers of the Danubean countries and other users requesting fast information obtain data and forecasts in the form of telex reports within 2 hours after forecasts have been made.

Forecasts are broadcasted for public use in the Hungarian Radio (at 13:45 in Hungarian, at 00:10 in Russian and French, and 00:30 in Hungarian; all in local times). Data and forecasts are also reported for subscribers in the form of Daily Hydrological Bulletin/Map, published regularly since 1895.

For more frequent information dissemination an automatic retrieval system, HYDROINFORM, was developed at VITUKI. The system is consisting of a micro-computer EMG-666/B, a telex interface and a 16 Kbyte external RAM. The telex requests of users are channeled directly into the HYDROINFORM that responds automatically and sends the required information to the users. Similarly to the TELEXDAT, the HYDROINFORM System can also be reached through the telex network. The TELEXDAT System is available as a HOMS Component (identifier: J90.2.05).

SUMMARY AND CONCLUSIONS

The paper overviewed the computerized operational forecasting system developed at the Hungarian National Hydrological Forecasting Service. The data collection and communication system is telexbased and partly automatic through the TELEXDAT system. Data processing and forecasting are performed on a 64 Kbyte computer in an interactive mode. This man-machine interaction enables to utilize the speed of computer as well as the experience of the forecaster. Two numerical models are run in parallel as decision aids for the forecaster. The first model (DLCM) is based on the adequate discretization of the continuous Kalinin--Milyukov-Nash cascade. The residuals of the discrete deterministic model are described by ARMA models of varying order. The discrete cascade and the time series model are than put up into a joint structural-stochastic model whose forecasts are recursively updated by the linear Kalman-filter algorithms. The second model that acts as a back-up is an entirely stochastic time series model based upon the adaptive self-tuning principle.

In summary: It is felt that models and microcomputers are excellent decision aids for the operational forecaster. Yet, the experiences of the forecaster should never be excluded from the process of hydrological forecasting.

REFERENCES

Ambrus, S.Z. and Szöllősi-Nagy, A. (1981a) A comparison between the efficiencies of real-time adaptive and traditional forecasting methods on River Danube, Int. Conf. on Numerical Modeling of River, Channel and Overland flow for Water Resources and Environmental Applications, IAHR, Bratislava, Section 1.2

Bartha, P. and Szöllősi-Nagy, A. (1982) The research and development program of hydrological forecasting in Hungary, in Hungarian, Vizügyi Közlemények Vol. 64, No. 1., pp. 7-30.

Ganendra, T. (1980) A Self-Tuning Predictor Applied to River Flow Forecasting, IIASA/WMO Workshop on Real Time Forecasting/Control of Water Resource Systems, October, 1976, Laxenburg, Austria, Edited by E.F. Wood and A. Szöllősi-Nagy, Pergamon Press, pp. 139-151.

Futaki, K. and Vargay, Z. (1986) Primary processing of hydrological data on microcomputers, this conference

Kalinin, G.P. and Milyukov, P.I. (1958) Priblezhennyi raschet neustanovivshegosya dvizheniya vodnykh mass (Approximate calculation of unsteady flow of water masses), in Russian, Trudy TsIP, No.66, Leningrad

Klemes, V. (1978) Physically based stochastic hydrologic analysis, in V.T. Chow (Editor) Advances in Hydrosciences,Vol. 11., pp. 285-356, Academic Press, New York

Nash, J.E. (1957) The form of the instantaneous unit hydrograph, Proc. IASH Assem. Gén. 3:114-131, Toronto, Ont.

O'Connel, P.E. and Clarke, R.T. (1981) Adaptive hydrological forecasting - a review, Hydr. Sci. Bulletin, Vol., 26., No. 2, pp. 179-205

Szesztay, K. (1957) Continuous daily stage/discharge forecasting on the Hungarian reach of the Danube, in Hungarian, VITUKI Beszámoló No. 6., Budapest

Szöllősi-Nagy, A. (1982) The discretization of the continuous linear cascade by means of state space analysis, Journal of Hydrology, Vd. 58, No. 1-2. pp. 223-236

Szöllősi-Nagy, A., Bartha, P. and Harkányi, K. (1983) Microcomputer based operational hydrological forecasting system for River Danube, WMO Techn. Conf. on Mitigation of Natural Hazards through Real-time Data Collection Systems and Hydrological Forecasting, Sacramento, Calif.

Wood, E.F. and Szöllősi-Nagy, A. (Editors 1980) Real-time Forecasting/Control of Water Resource System, Pergamon Press, Oxford.

FLOOD FORECASTING FOR A LARGER CATCHMENT USING RECORDS FOR SMALLER CATCHMENTS

D E Goring[1] R P Ibbit[2] M E U Taylor[3]

The Flood Forecasting procedure described here was developed to provide flood forecasts for catchments with a short response time which are occasionally subject to heavy rain, and where flow information is available for only about half of the tributary inflows.

This situation is often met in New Zealand. The country is often fairly steep and sparsely populated, access is difficult, and economics prevent measurement stations being installed on all tributaries. Nevertheless, effective flood-warning systems are necessary to protect settlements at the downstream outlets of the catchments.

Often the period of record is relatively short, sometimes less than ten years, and so far it has not been possible to establish good correlations between rainfall and flow.

BACKGROUND

Two classes of flood routing techniques may be distinguished. Hydraulic methods involve the solution of the St Venant equations or approximating these. Hydrological methods use a system or "black box" approach.

The hydraulic method had two practical drawbacks for the problem under consideration. One is the necessity to have some relationship between flow and stage, which was not always available in the catchments concerned. More seriously it is difficult, if not impossible, to account for ungauged tributary flow. This is a major problem for a sparsely populated country like New Zealand where, although most of the major rivers have records, few tributaries do.

It was therefore decided to use a hydrological method which treated the river system as a "black box" in which the output and some of the inputs could be measured, and to derive a relationship between these, the "Transfer Function".

1,2 Scientist, Hydrology Centre, Ministry of Works and Development, Box 1479, Christchurch, NEW ZEALAND

3 Research Director, Water and Soil Directorate, Ministry of Works and Development, Head Office, Box 12 041, Wellington North, NEW ZEALAND.

World Meteorological Organization (ed.), Microprocessors in Operational Hydrology, 179-182.

For convenience in calculation a linear systems analysis procedure was preferred. However, because flood wave propagation is unlikely to be a linear process, an analysis of the errors involved in using a linear instead of a non-linear model was made.

This showed (Goring, 1984) that although the linear model propagated the crest at 0.58 times the velocity of the non-linear models and steepened the back rather than the front of the wave, it maintained the correct crest flow and the difference in shape after propagating 160 km was so small that a linear model which adjusts for the error in propagation velocity should be satisfactory.

Originally the derivation of the "Transfer Function", which describes the relationship between the measured tributary inputs to the river system and the downstream discharge, required more powerful computational capabilities than were available on a microcomputer and therefore had to be carried out on a main frame computer.

Microcomputer procedures have now been developed, so that all stages can be carried out by microcomputer. Once the transfer function has been derived it is readily usable on a microcomputer to provide rapid and easily updatable forecasts.

The derivation of the transfer function uses Fourier transforms involving harmonic analysis based on convolution theory (Goring, 1984). Other hydrological routing techniques such as storage routing and the Muskingum method can be shown to be special cases of the more general linear systems technique in which the transfer function has a predetermined form.

PRINCIPLE OF APPLICATIONS

The transfer function method has been developed to provide flood forecasts in situations where flow measurements are only available for some of the tributary catchments (Fig 1).

It is obvious that rainfall events approaching the catchment from different directions will give rise to different sequences of flow events in the various tributaries. It may therefore be necessary to derive several different transfer functions for the catchment, depending upon its response pattern.

Obviously the more data available the better the estimates which can be made of the transfer functions. Over a number of implementations it has been found that 10 years of records produce enough large floods to satisfactorily identify the transfer functions. In general, use of any more than six floods does not change the transfer functions.

This procedure gives a good estimate of the time of arrival of the flood peak and the shape of the hydrograph.

Application of the current downstream flow to the forecast allows the estimated hydrograph to be shifted linearly up or down to fit this flow. This enhances the accuracy of the forecast by accommodating the real time conditions at the downstream site (Fig 2).

The procedure has been operational on two rivers for 13 months and has been applied to five other rivers in the last year. The package is written in Fortran IV and is used on a range of micro- and mini-computers. Operation of the forecasting system can be successfully carried out by hydrological personnel unskilled in the operation of computers. However, determination of the transfer function must be carried out by skilled staff.

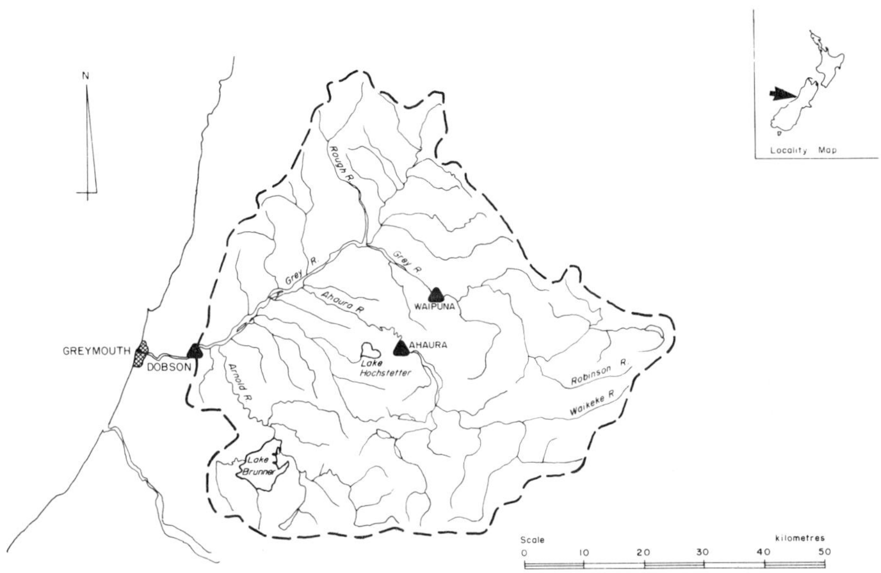

Figure 1: *The Grey Catchment comprises a number of tributaries flowing into the Grey River. Radio telemetry stations are located at Waipuna and Ahaura. The measured tributary inflows arise from only 49% of the catchment area. A verification station is located at Dobson.*

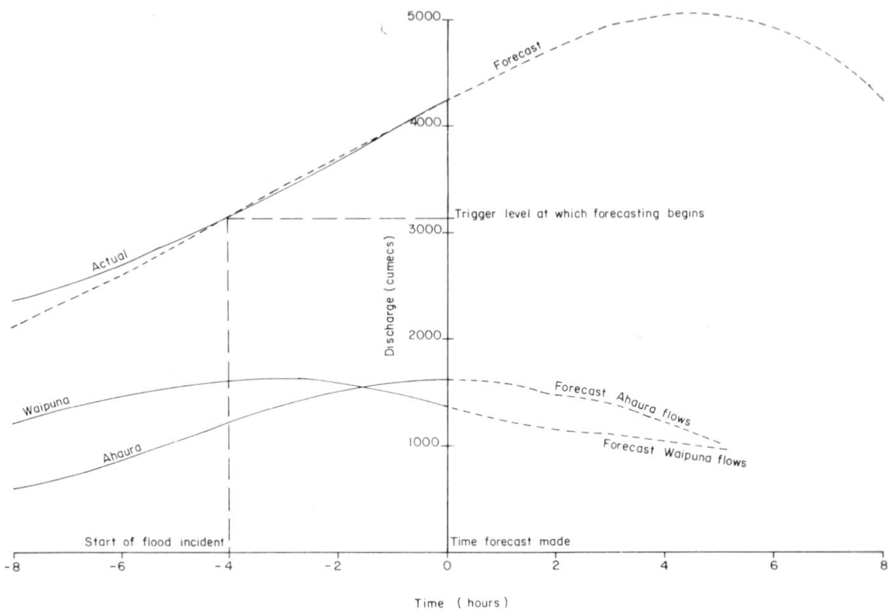

Figure 2: *A typical microcomputer output during a flood forecasting session. Forecasting began at time -4 hrs when the flow exceeded the "trigger level" of 3000 cumecs. Telemetry stations at Waipuna and Ahaura provided signals used to forecast the flow at Dobson. This flow peaked at 5000 cumecs, 5 hours after the forecast was made, which corresponds to a flood level of 6256 mm.*

THE USE OF MICROCOMPUTERS IN OPERATIONAL HYDROLOGY IN JAPAN

F. Yoshino*

1. The Current State of the Application of Microcomputers to Flood Prediction in Japan

Of the flood prediction systems currently used in 109 water systems, we examined those in which microcomputers are used. The results of our investigation are;

(a) With regard to calculation technique, almost all of the systems adopt the correlation methods such as the R-H and H-H methods, and the storage function method. Many of them use both techniques.

(b) The storage function system is utilized for 84 rivers, with 42 rivers having large computers, while the remaining rivers adopt microcomputers.

(c) In 2 rivers the storage function and the unsteady flow estimation (unsteady channel) technique is applied.

(d) As to data input, approximately 40% of the storage function systems adopt on-line systems.

(e) With respect to necessary time for calculation 69% of the storage function systems perform estimation within 15 minutes, yet take a longer time for rivers with complex models.

Table 1. Flood prediction estimation techniques of systems using microcomputers

Method	Systems
Storage function method	28 systems
Correlation method(H-H, R-Q-H, etc.)	7 systems
Kalman filter method	1 system
Unit-hydrograph method	1 system
Tank model method	2 systems
Combination of more than two of the above-mentioned methods	13 systems
Total	52 systems

Table 2. Data input methods

Method	Systems
On-line	15 systems
On-line (key board input)	37 systems

*Head of Hydrology Division, Public Works Research Institute, Ministry of Construction, Japan.

World Meteorological Organization (ed.), Microprocessors in Operational Hydrology, 183-193.

Table 3. Accuracy of water-level prediction of the systems

Within	30cm	13 cases
Within	50cm	17 cases
Within	100cm	9 cases
More than	100cm	4 cases

2. Example of Flood Prediction by Microcomputer

The future water-storage and discharge have been estimated conventionally through runoff systems based on hydraulic and hydrological data for flood prediction, and the estimated value inevitably contains an error relative to a measured value. This error mainly stems from the approximation of a physical model to an actual phenomenon, errors in rainfall estimation, and errors in the water-level/discharge transformation. For this reason a feedback system to converge the estimated value to the measured value is required.

In this report, the Yodo River is examined as an example of the application of a feedback system to flood prediction using microcomputers. In practice, a system using a Kalman filter, which has become a powerful tool in flood prediction, has been developed. In addition to this microcomputer system, the Yodo River has adopted a system in which rainfall prediction results obtained through a radar raingauge are input to a large computer. Thus, the system using a microcomputer is a supplement to the larger system.

2-1. Flood prediction model

This model estimates the discharge at each watershed end (Katsura, Nadokoro and Kamo) through the storage function method by inputting the rainfall at the sites representing each watershed, branch water-storage and discharge at dams into a Y.H.P. 9845B computer.

Then, channel unsteady flow estimation is performed by treating the values at Nadokoro and Kamo, and the respectively discharges from the Amagase Dam as lateral inflows and inflows at the upstream end, to predict water level and discharge at each site.

(1) Reading of initial hydrological data

Railfall and discharge data for 6 hours before the starting time of flood prediction are read.

(2) Basin yield estimation (estimation of initial discharge)

Discharge at the Yodo River basin is calculated from the average basin rainfall obtained from rainfall at the observatory station using the storage function method.
The constants for the storage function were set as follows, after the analysis of the constants;

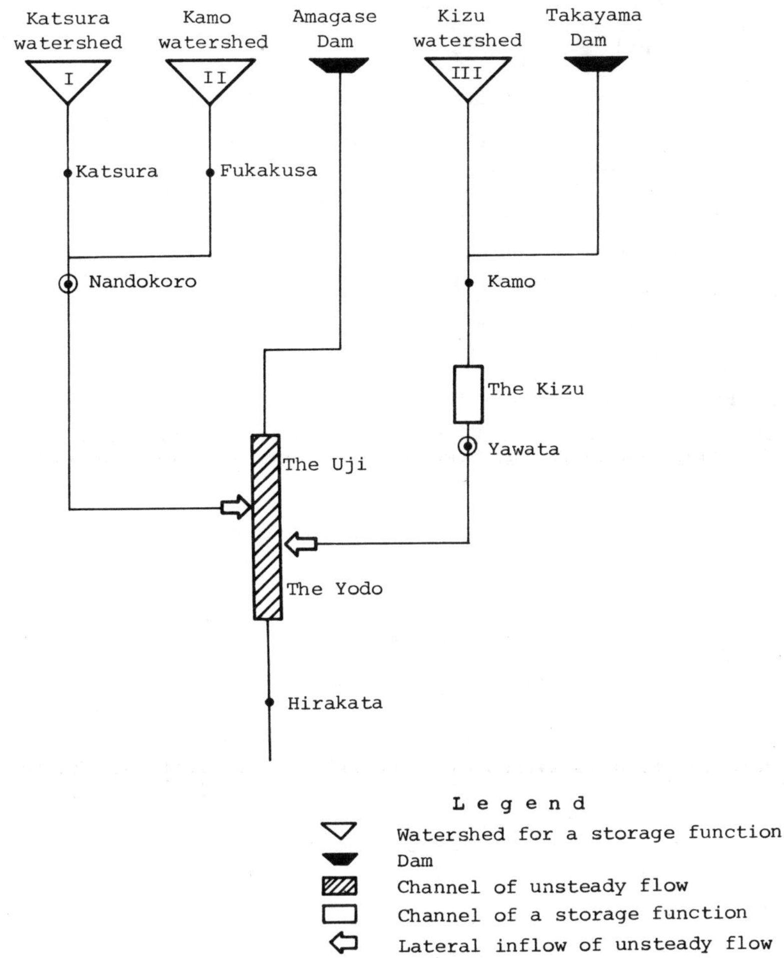

Figure 1. Conceptual diagram of the prediction model

Table 4. Watershed constants

Watershed \ Items	Base point	$A (Km^2)$	K	P	$T\ell$ (Hr)
The Katsura	Katsura	895.0	45.0	0.452	4.8
The Kamo	Fukakusa	187.0	48.3	0.533	1.3
The Kizu	Kamo	854.8	45.0	0.374	5.5

For the estimation of the effective rainfall, the runoff-parameter (F) is used to approximate the estimated discharge to the measured discharge based on the relation with the feed-back system mentioned later, instead of using the primary runoff-ratio.

(3) Reading of the prediction time length and the predicted rainfall at each hour.

Predicted rainfall and dam discharge are set by input of the measured values.

(4) Reading of actual hydrological data at each hour.

(5) Estimation of watershed runoff

The runoff estimation is carried out in the same way as (2), but a feedback system is introduced to bring the predicted value close to the measured value, since there is a difference between them.

i) Setting of initial conditions

When the predicted value differs from the measured value at the prediction hour, runoff estimation is performed by replacing the predicted value with the unit runoff corresponding to the measured value. (Figure 2)

ii) Correcting by the runoff-parameter (F)

The errors between the predicted value and the measured value produced from the errors in effective rainfall estimation, are corrected through the run-off-parameter (F). The runoff-parameter is defined by the following equation :

$$R_{ave} - q_e = \frac{dS_e}{d_t}$$

$$S\ell = K.Q_s{}^p$$

$$Q\ell = \frac{1}{3.6} A \cdot q_e \cdot F$$

R_{ave}: rainfall, q_ℓ: unit runoff, S_ℓ: unit storage, K and P: constants, and Q_ℓ: runoff

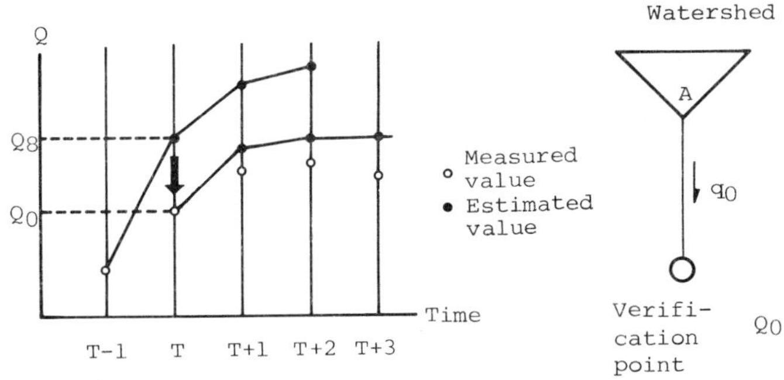

Figure 2

(6) Estimation of channel unsteady flow

Tracking of flood in the Uji is carried out through the estimation of unsteady flow.

i) River channel divide and time differentiation

As shown in Figure 3, the river channel is divided into Δx=4,000m units (distance differentiation), and flood tracking estimation is then performed under the condition of Δt=3 minutes (time differentiation).

ii) Boundary condition

The boundary point at the u stream end is set at 51.0km cross section (The Uji), and the discharge from the Amagase Dam is given. In addition, the predicted discharge hydrographs at Yawata (The Kizu) and Nandokoro (The Katsura) are given to the channel sections between No. 3 and No. 4, and between No. 4 and No. 5 as latent discharges.

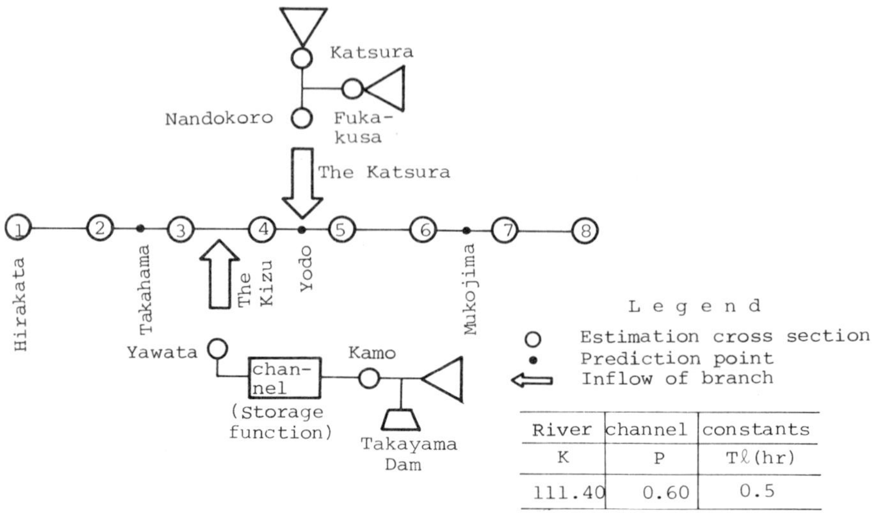

River	channel	constants
K	P	Tℓ(hr)
111.40	0.60	0.5

Figure 3. Schematic diagram of the river channel

2-2. Verification of the flood prediction model

An example of the results of the flood prediction estimation for the flood in August, 1982 and the flood in September, 1976.

The prediction hour is 8 hours, and the prediction type is 1.

The o mark in the figure denotes the actual value.

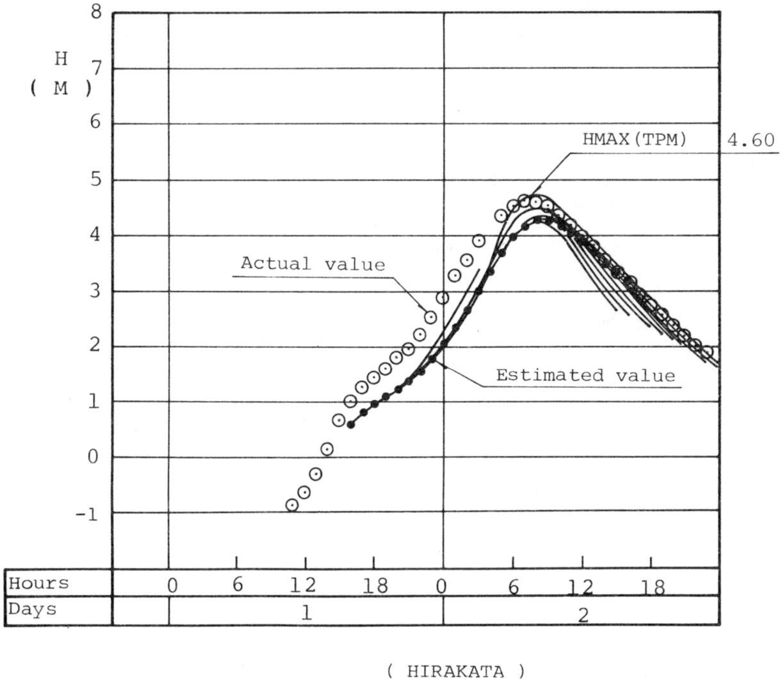

Figure 4. Result of prediction estimation (flood in August 1982)

3. Data Processing Using IC Memory

Since the recording system of the self recording flood-meter and self recording raingauge conventionally used was of the analog data type using a pen recording system, it took much time and manpower to read the hydrological analyses based on the data.

In order to improve this shortcoming, the application of LSI semiconductors has become more common.

For this reason, we compared the LSI methods including the traditional recording methods, to examine the use of the new-type IC memory. Table 5 is a summary of the results. The portable memory for the recorders introduced here is called the digital memory pack (D.M.P.).

Table 5

No.	Item to be compared \ Method	DMP method	Self-recording method	Telemeter method
1	Sampling interval and data density	Intermittent data with at least a 10 minutes interval	Continuous	Usually 1 hour and minimum 10 minutes
2	Data recording capacity	1∿3 months (10 minutes sampling, 16K bit∿64K bit)	1∿3 months recording paper	List and record every data collection at the data management center
3	Direct reading and confirmation of data	Direct confirmation by eye is impossible	Possible by the use of a checker	Confirmed by eyes and conversion
4	Data retrieve method	Visit the site and exchange MDP	Visit the site and exchange recording paper	Collect at the data management center all the time
5	Errors in data reading	none	According to person errors in reading	none
6	Easiness in data processing	Direct processing by CPU is possible	Manual reading and listing from recording paper	Partly direct data processing by CPU

3-1. Outline of the D.M.P. system

The configuration of the D.M.P. system is shown in Figure 5.

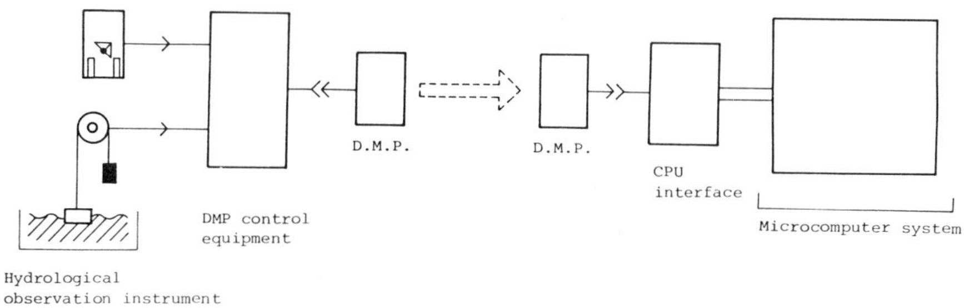

Figure 5. The composition of the D.M.P. system

(1) Hydrological observation instrument -- Instrument in which measured values are put out through electric signals.

(2) OMP control equipment -- Measured signals are printed into DMP.

(3) D.M.P. -- See 3-2.

(4) CPU interface -- Connect D.M.P. with microcomputer.

(5) Microcomputer system -- Processing of D.M.P.-recorded values and analysis of hydrological data.

3-2. Characteristics of recording media used in DMP

IC memory and bubble memory are selected for comparison. Table 6 summarizes the results.

Table 6. Comparison of several kinds of DMP recording methods

Memory type	Data storage	Environment for use	Electricity consumption per 1 bit (about)	Reuse limitations	Cost per bit (about)	Operatability	Capacity per bit	Wide use as memory cell	Note
IC memory, C-MOS S-RAM	Battery support is necessary due to volatile memory.	-20°C∿40°C	During operation μw/bit Stand-by μw/bit 5V single source (16Kbit/chip)	No problem	16Kbit/chip 1,000 Yen 0.06Yen/bit	To place in a metal case is necessary. Slightly difficult.	16Kbit/chip	○	
IC memory EEP ROM	"	-20°C∿40°C	During operation μw/bit Stand-by 0 (additional +30V is needed) During operation μw/bit Stand-by 0 (5V single source) (16Kbit/chip)		Expensive, 16Kbit/chip 10,000Yen/unit 0.06Yen/bit	"	16Kbit/chip	×	
Bubble memory	"	0°C ∿50°C From-10°C to 0°C preheating is required.	During operation μw Stand-by 0 (1Mbit/cassette)	No problem	Bubble cassette 4∿50,000Yen/unit 0.2Yen/bit	Excellent operatability because of cassette for exclusive use	128Kbite/cassette	△	

Currently, in Japan the application of C-MOS-S-RAM memory to the recording part of rain-gauges is becoming more popular. Also, systems using bubble memory have been developed for practical use, and the further use of these is widely anticipated in the processing of hydrological observation data.

A COMPREHENSIVE HYDROLOGICAL DATA SYSTEM

M.E.U. Taylor[1] S.M. Thompson[2]

INTRODUCTION

The water resource is very important to the New Zealand economy for hydropower, irrigation, etc. On the other side of the coin, floods severe enough to cause damage to property occur somewhere in New Zealand about once a year on average. Accurate water resource data is therefore given a high priority by the water resource management authority.

Some of this information is collected by the 20 Catchment Authorities who have regional responsibility for flood forecasting, protection, etc. The bulk of the rest (for power development, irrigation supply, scientific research, etc) is collected by Water and Soil Directorate of Ministry of Works and Development.

In the mid 1970's a scheme for telemetering much of this data was proposed. Feasibility studies showed that in addition to improved flood warning and forecasting, substantial operational savings could result in hydrological data collection. Many of the key hydrological and rainfall measuring stations are located in inaccessible country requiring access by helicopter or a several-day trip using four wheel drive vehicles.

A survey of the telemetry systems which were then commercially available showed that, although there were many eminently suitable for flood warning purposes, none met the criteria of robustness and accuracy essential for a hydrological data system.

An appropriate specification was jointly worked out by the Catchment Authorities and Ministry of Works and Development. It was decided to develop a suitable telemetry system by a combination of development contracts and "in house" development.

The commencement of the development contract was delayed for three years by problems in getting approvals but eventually commenced early in 1982.

THE OVERALL SYSTEM

The overall system provides accurate measurement, data transfer, quality assurance, data handling and storage, and applications software (including flood forecasting).

---

1 Research Director, Water and Soil Directorate, Ministry of Works and Development, Box 12 041, Wellington North, New Zealand

2 Scientist in Charge, Hydrology Centre, Ministry of Works and Development, Box 1479, Christchurch , New Zealand.

World Meteorological Organization (ed.), Microprocessors in Operational Hydrology, 195-203.

SENSORS

Stage

Initially the telemetry equipment has been interfaced to existing punched tape recorders. These recorders can be accurate to one millimetre and it is difficult to persuade people to accept less accuracy when telemeters are installed.

A pressure sensor has been developed which is as accurate as a recorder with a float in a stilling well. Each sensor has an individual laboratory calibration, which compensates for temperature effects at both the sensor and microprocessor and is stored in read only memory (ROM). The effects of surface waves are overcome by digital averaging so that the installation costs of a stilling well are avoided. This sensor's electronics use the telemetry's power supply and can fit into the telemetry's hermetically sealed container or function as a plug in module. High accuracy is attainable (Figures 1 and 2). Use of pressure sensors enables considerable economies to be made in constructing a hydrological measurement site.

Rainfall

The telemeter can be interfaced to commercially available tipping bucket raingauges. This provides the requisite accuracy and robustness. There is therefore no requirement for further developments.

Water Quality

Satisfactorily stable and accurate temperature and conductivity sensors are commercially available. These can be interfaced to the telemeter. Available commercial dissolved oxygen, pH and turbidity sensors are not sufficiently stable for prolonged routine use. Investigation of a possible route to a stable dissolved oxygen electrode is currently producing promising results.

RADIO TELEMETRY EQUIPMENT (Figures 3 and 4)

The radio telemetry equipment which has been developed consists of four kinds of unit:

(a) A base unit in the field office which asks remote recorders by radio to transmit their data then receives this data and passes it to a microcomputer.

(b) Remote units at each field recorder which store up to eight days data in random access memory (RAM) and maintain a low power radio listening watch ready to power up and transmit data when asked.

(c) A support unit, normally carried in a back pack, which may be plugged into the hermetically sealed remote units when they are visited to check, or modify, the parameters governing data storage, and to recover data when the radio circuit is not operational.

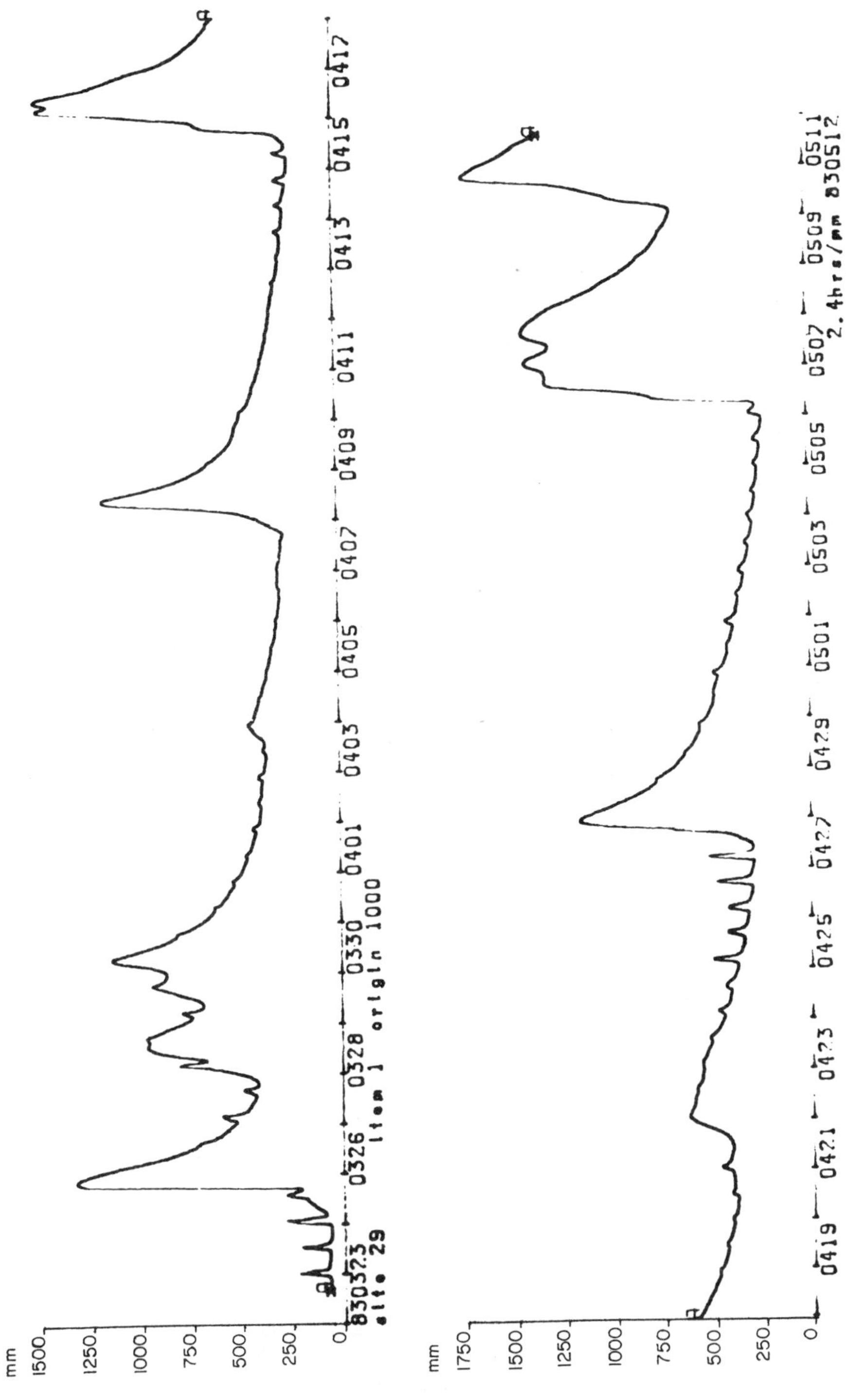

Figure 1: Float record

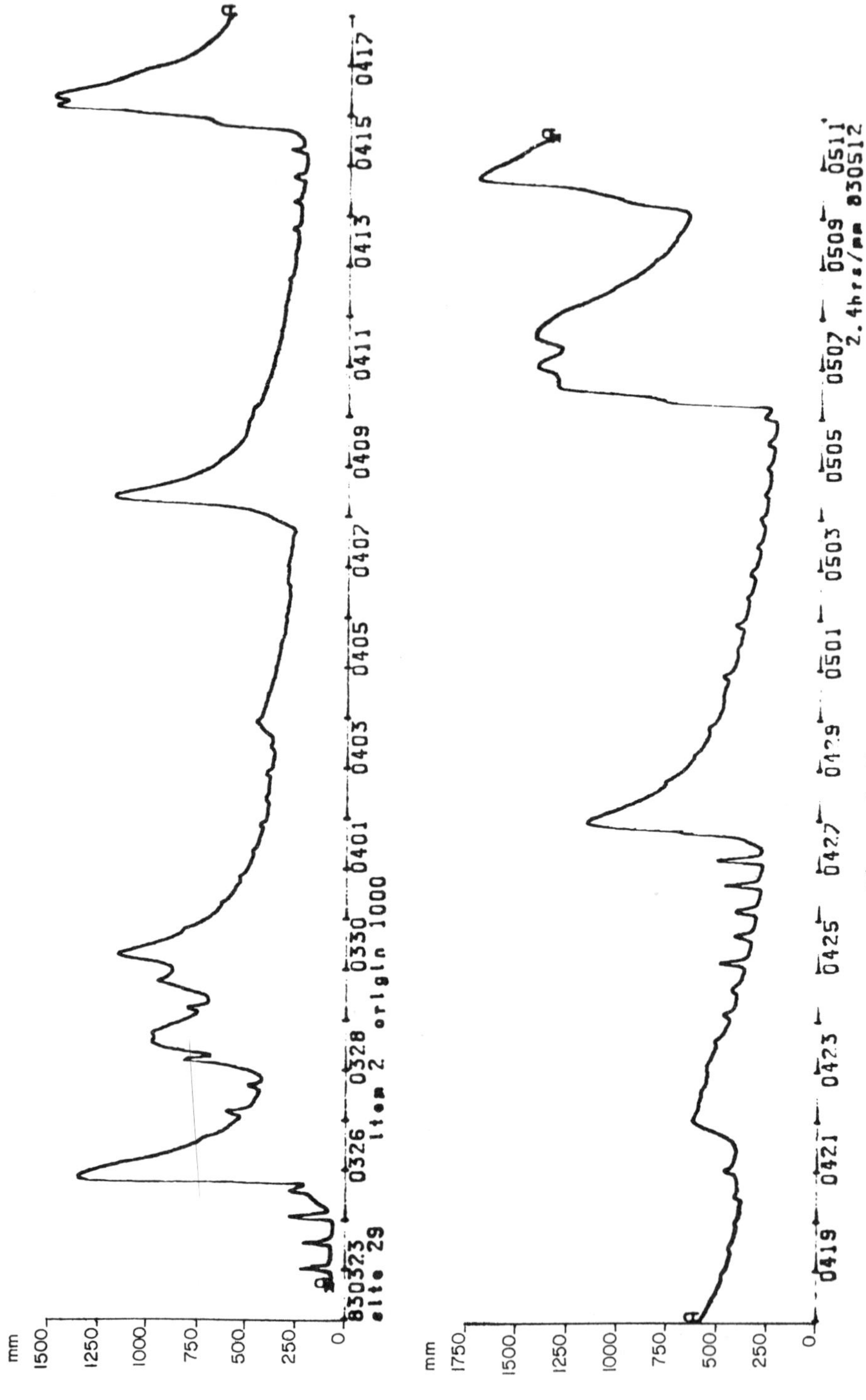

Figure 2: Pressure transducer record

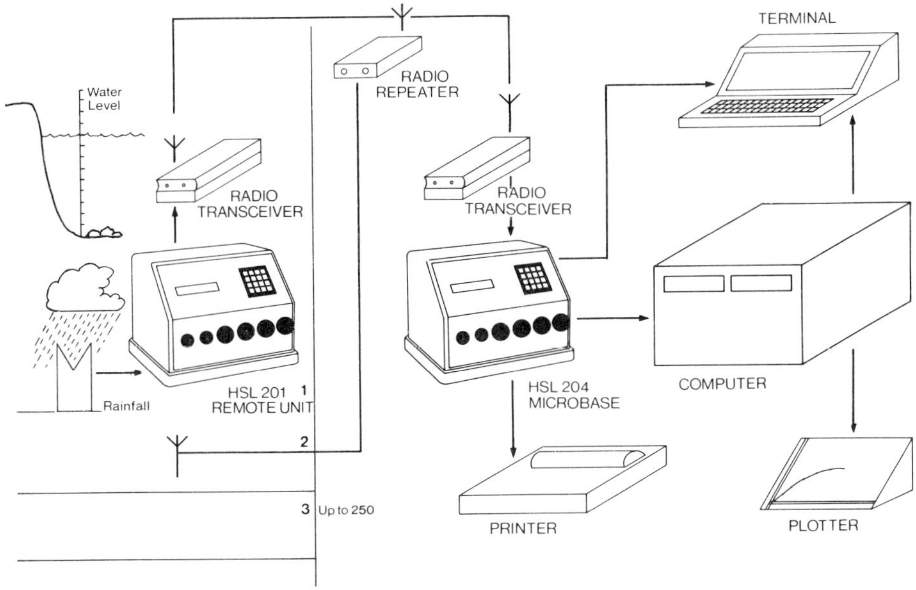

Figure 3: System diagram.

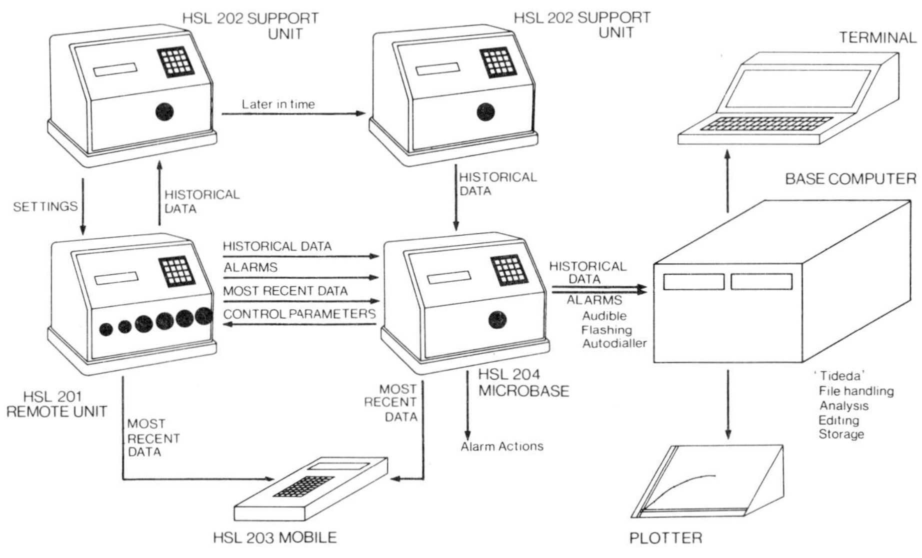

Figure 4: Data flow diagram.

(d) A mobile unit, normally mounted in a vehicle, which uses radio to recover data and the alarm status of any remote.

The system has been designed for operation on existing VHF voice radio networks. Initial development also looked at communications satellites, which are technically very satisfactory. However, concern was felt about committing a key data transmission network to a system over which New Zealand had no direct control and where there was an element of uncertainty regarding guaranteed access and pricing arrangements. This doubt was reinforced when the experimental satellite communication system was closed down by the satellite operator at very short notice.

The VHF system has been designed to transmit digital data successfully over noisy voice networks. A high degree of data security against loss is provided. The remote telemeter is hermetically sealed, immersion resistant, exceptionally robust and retains a record of the data for up to eight days before this is overwritten. Failure to receive a data transmission does not therefore result in loss of data. Precautions are taken against unauthorised interference, power loss, etc. Alarms can be raised when signals exceed preset limits, providing warning against battery malfunction, sensor malfunction, flooding, etc.

Every transaction (data transfer, operating system adjustment, alarms, etc) is logged. Every set of readings is individually identified and the time recorded on a data header.

To minimise unauthorised interference, the remote telemeter can only be reset from a designated microbase or from a special support unit. However, it can be interrogated manually on site or remotely from any microbase or mobile unit which operates on the appropriate wavelength. This facility enables the Water and Soil Directorate and the Catchment Authority networks to be interfaced.

The radio transceivers share a VHF channel with voice traffic, and therefore use existing repeaters. The microbase automatically waits until the radio channel is free, then transmits its request to, and receives data from, the remotes. Routine data collection is scheduled during the night when other radio traffic is least, although it may be at any time. It requires 30 seconds of radio transmission to recover eight days of 15 minute values from two sensors.

Data is stored at a preset interval of 7½, 15, 30, 60, 120 or 240 minutes and transmitted at preset times or on interrogation, for example, after a flood alarm has been raised.

Power consumption is extremely low. A standby supply guards against loss of record.

The remote unit can be fitted with control systems to operate flood gates, pumps, irrigation systems, automatic samplers, sensor calibration systems, etc.

A microbase provides the specialist telemetry assistance to enable the base computer or telemetry operator to retrieve data and control the system. It will activate alarms and automatically dial up to eight preselected telephone numbers until the alarm is acknowledged. It provides the system log and controls communication between the remote telemeter and the support unit or mobile unit.

DATA HANDLING

The microbase communication to a base station computer. In the Ministry of Works and Development system the base computer uses microTIDEDA to process and store the data. Automatic quality assurance checks can be carried out, as described in the paper on microTIDEDA in this seminar.

OPERATIONAL EXPERIENCE

The telemetry system is currently operating at three base stations, controlling some eight remote units. Installation is in progress at one further Ministry of Works and Development base station and three Catchment Authorities. It is planned to have sixteen Ministry of Works and Development base stations and some 80 remote sites operational by March 1985.

An example of a telemetered record from a tidal river is shown in Figure 5.

A number of minor commissioning problems were experienced when the telemetry system was first tested. At the time of writing these all appear to have been resolved except for a minor software problem with the rainfall recorder.

Fifteen pressure sensors are presently being field tested. They are functioning satisfactorily.

The microTIDEDA data handling system is fully operational and field tested in 16 Ministry of Works and Development stations. It has received enthusiastic acceptance from users.

The flood forecasting software is in operation in four Catchment Authorities which have installed the necessary telemetry. It is fairly robust in use. On one occasion it was successfully used by a Catchment Authority's accountant when the Chief Engineer and Water Resources Officer were both absent from the office. Nevertheless,it is recommended that it be operated by appropriately trained technical personnel.

CONCLUSION

Introduction of a telemetric hydrological data system is greatly enhancing New Zealand's hydrological survey. Whereas earlier all data editing involved handling both by a centralised data processing system and by later manual editing by field staff, nowadays the data entry EDP is all carried out by field staff.

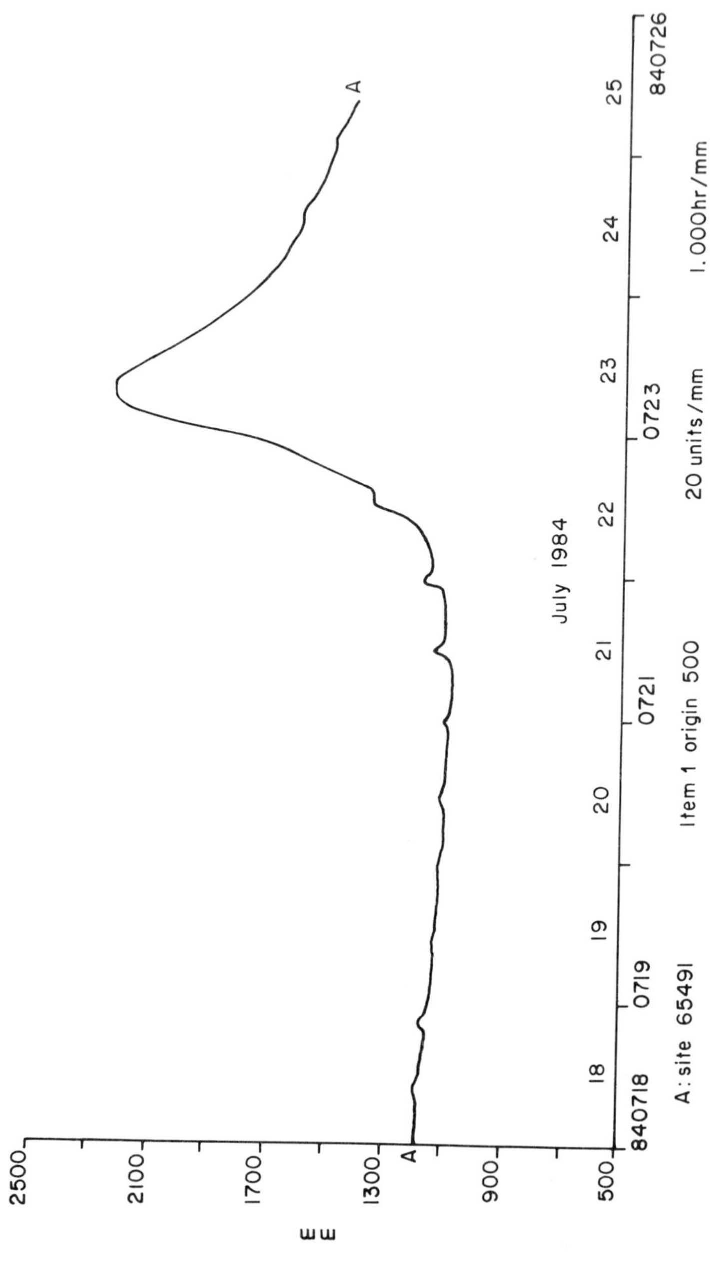

Figure 5: One-week record of telemetered data, Waimakariri River, New Zealand

Time savings amount to weeks and sometimes months, and editing is now done while events are fresh in staffs' memories. Job satisfaction has increased enormously.

When the telemetry and integration of the whole system is complete in four years time, the automated data quality assurance procedures will assist greatly in standardising the quality of data produced by the 22 different national and local agencies involved.

Substantial savings are expected when full conversion to the system is completed. The New Zealand Treasury estimates the return from operational savings alone would be 24% pa. The savings resulting from better flood forecasts are an additional bonus.

EXPERIENCE ON THE USE OF AN AUTOMATIC DATA COLLECTION AND TRANSMISSION SYSTEM IN A HYDROLOGICAL RESEARCH BASIN

F. Koch and H. Lang*

INTRODUCTION

The hydrological research basin Rietholzbach is a typical pre-Alpine area in the northeastern part of Switzerland. It serves to investigate different problems: the spatial and temporal variations of rainfall-runoff processes; the various methods of determination of evapotranspiration for a point and for catchment areas; the information content of a small scale basin with respect to the large scale river system of Thur and Rhine, for which daily operational runoff forecasts are issued. This paper describes the automatic data collection system of the small research basin and the experiences so far gained.

The instrumentation in the Rietholzbach catchment (3.18 km^2, 680-950 m a.s.l.) was started in 1975. It was decided to use an automatic data collection and transmission system to have a continuous control of the measuring system and of the sensors used in order a) to reduce time for maintenance b) to reduce loss of records in case of failures in the observation system c) to have the data available at any time for processing in the operational real-time forecasting and in the analysis of the actual hydrological processes.

The data collection system measures at present the following components:

- Discharge gauging station Rietholzbach, including a pneumatic gauge for analogue and digital recording.
- Hydrometeorological station Büel, including different sensors for airtemperature, air humidity, soil temperature, precipitation, radiation components, lysimeter weight and outflow (3.14 m^2 surface area, 2.30 m soil profile), gamma-radiation device for snow water equivalent and soil moisture (also used for class-A-pan water level), meltwater lysimeter.

There are principally two types of electronic measuring systems: One uses a changeable data carrier (magnetic tape etc.). The other transmits the data directly, or after intermediate storage, by way of an electric connection (wire, or wireless) to the point of use. In this case the distance between the catchment basin and the central institute in Zurich is about 60 km. The field stations are connected to the public telephone network and to power supply. With view to the requirements and to the approx. 1800 values per day of records an efficient and reliable data collection and transmission system appeared to be necessary. To meet the case of failure and for comparison purposes some of the components had to be recorded by standard mechanical instrumentation (air temperature, humidity, wind vector, precipitation, potential evaporation, and discharge).

DESCRIPTION OF THE ELECTRONIC DATA COLLECTION AND TRANSMISSION SYSTEM

The system consists of a central unit located at the institute residence in Zurich, and at present of two field stations in the research basin Rietholzbach: one is recording water level and temperature at the discharge gaging station, the other field station is recording the signals of 20 sensors at the hydrometeorological station Büel (Fig. 1). Both units are adjusted to 5-minute intervals; for some of the parameters 1-hour intervals are selected for the interrogation.

*Federal Institute of Technology, Zürich

World Meteorological Organization (ed.), Microprocessors in Operational Hydrology, 205-212.

The first step to record a non-electrical parameter is to transform it into a reproducible electrical signal of high accuracy. For its further processing it has to be changed into digital form. The accuracy requirements need a corresponding amount of effort in the electronics. The present equipment uses an 11-bit analogue-digital unit, which provides a resolution into 1023 steps and a corresponding accuracy in temperature of 0.024 ^{o}C per bit (range -20 ^{o}C to +30 ^{o}C). Together with the input signal amplifier the final accuracy in the temperature measurement amounts to 0.1 ^{o}C.

The field stations consist of the following components (Fig. 2):

- Sensor: Transformation of the physical signal into an analogue electrical signal.

- Input signal amplifier: with data input module: The amplifier adapts the different electrical input signals to a unified output voltage of 0...10 V. The data input module connects the input signal amplifier to the microprocessor (µP)

- Analogue - digital unit (A/D-unit): This unit transduces the measured analogue signals of the amplifier (0...10 V) into 11-bit-binary values.

- Central processing unit (CPU): This unit contains a microprocessor (MC 6800) and the timer with the following assignments: Control of the data input modules - Data reduction (averaging) - Determination of extreme values - Control of the display - Control of the data transmission - Decoding the return signal.

The programs used by the CPU are standard versions, adapted to the specific requirements. The whole program consists of a member of subprograms, which are stored in a read-only-memory (EPROM). The program comprises approx. 8 k-Byts.

- Random access memory (RAM): This memory is set up in C-MOStechniques. It can be extended up to 32 k-Byts. The memory is preserved in case of power failure by an additional battery.

- Real time unit: This unit uses a quartz-component at a base frequency of 1.048 kHz, from which all other frequencies in the system are derived (except the CPU frequency). It is used for timing the data input module and as interrupt signal for the microprocessor. This unit contains a meter for minutes, hours, and days for manual adjustment.

- Display module: It serves as connection between the microprocessor and the data display. This unit allows to display the measured value of each sensor, and the time. The display is based on a 6-decade-LCD.

- Modulation-demodulation-unit: This unit represents the interface between microprocessor and telephone system. It provides the connection between the databus and the modem. All signals from and to the modem are galvanicly disconnected by opto-isolators. The control processes of the modem are determined by the program.

- Power supply including back up: The power supply module provides the adequate voltage to each of the units. In case of failure the power supply is provided by a back up battery which ensures power supply over a period of up to 24 hours.

The interrogative interval of each of the sensors can be selected individually. The time signals (minute-impulses) sent by the timer are analyzed by the data input module, and, corresponding to the program, an interrogation cycle may be started. The sensor then is interrogated, the obtained signal is converted into digital form by the A/D - unit; some of the sensors already provide digital signals. The digital values together with the time signal, are then stored in the memory (RAM). The reading of the stored data is started by the central unit via telephone connection. There are two ways:

a) interrogation of the actual signal of all sensors

b) interrogation of the 24 hour data storage (it is possible to read the whole data of the past 3 days)

c) transmission of all the data by 1 to n intervals ($n < 30$)

d) transmission of the data of only one sensor.

The interrogation does not affect the storage in the memory, nor the interrogation cycles, and an arbitrary number of interrogations is possible. The data are provided by the field station via modem, which converts the digital signals into acoustic signals in the range of 300-3600 Hz; the mode of telephone transfer uses parallel transmission processing. This method allows to transmit 40 signals per second. It may be regarded as slow, but with the great advantage of low requirements of qualification with respect to the telephone transmission lines. The field station can be interrogated by the central unit only with an automatic dial-unit for data transmission. After the completion of the data transmission into the memory of the central unit, they can be put on various periphery devices (output form in ASCII). The central unit is taken as the main device of control, transmission and data output. It allows also the possibility of manual interference with the field station using the central unit, as for instance adjusting the time and the day number. Further processing of the data has to be done by computer facilities.

PRACTICAL EXPERIENCES IN THE DAILY USE

The two field stations are automatically interrogated once a day (at 6.00 a.m.). The transmitted data are put on a computer compatible punch-tape, which is read into a desk computer. Here the data (in units of voltage) are converted into the final units: degrees Celsius, humidity etc. A print-out and a graphical plot are immediately produced to allow a continuous check of the performance of the field stations, the sensors, and of the system as a whole (Fig. 3). At the same time this information provides quite a complete picture of the actual hydrologic situation in the field and is available for use in the operational forecasting. For the further processing the data are intermediately stored on magnetic tapes, and finally read into permanent files at the ETH computer centre. It is considered as an important point, that this system allows to detect any failures immediately and to minimize the corresponding data losses.

Maintenance

After having established the whole system, and after all the computer programs have been worked out to be ready for use, the maintenance of the system is regarded as relatively simple, as long as the electronic components do work.

Central unit: Failures of the central unit , can be discovered quickly and repaired in most cases, if trained people and adequate facilities are available. The three-day-storage-capacity of the field stations is of great importance in order to avoid or reduce loss of data while the central unit is under repair.

Field stations: Failures of the field units are in general immediately detected at the central station but the diagnosis has usually to be done at the site in the field. In fact the field units are frequently exposed to severe climatic conditions, which should be taken into account already in the design stage. A very strong disturbing factor are the electrical processes in the atmosphere, in particular lightnings usually connected with excess voltage. Solid state techniques can be badly affected even from small voltage pulses, causing damages of high costs. The voltage pulses can be caused by switching interference in the power supply networks, by static charges, or by lightnings.

In the discussion of this problem the costs and efforts for protection measures have to be taken into account. It seems to be necessary to include protection against excess voltage from nearby lightning activities and from electric charges of overhead power lines, while a full protection against direct lightning stroke seems practically impossible. In our case the protective measures consist of a gas lightning protection fuse with adjoined filter and transzorb diode. A further problem arises with extended networks, when ground loops are caused by a number of sensors, which are electrically connected to the ground. This should be avoided.

Failures

Since the start of the operation of this data collection system in 1975/76 there have been four serious lightning strokes, two of them causing great damages because of insufficient protection measures. The field units needed several weeks to be repaired. After the installation of the above mentioned efficient lightning protections, all following lightning-caused-failures were less serious and could be repaired in a short time and at the site without long interruptions of the records. A different cause of typical breakdowns are the occasional interruptions in the power supply. In particular in winter during snowfalls at temperatures near the freezing point, snow load formation on the power lines causing fractures of the wires. In these cases there is no loss of already recorded signals in the memories, but a temporary interruption in the continuous operation of the records, because the analogue unit is not provided by the battery back up.

Breakdowns due to failures of electronic components have occurred on the average once a year, mainly units of a first generation, using conventional electronics. In April 1982 the core of the system was replaced by a micro processor controlled unit. The following table shows the annual data losses in percent of the total during the last 9 years:

Year	Data losses	Remarks
1975	7 %	various reasons
1976	6 %	" "
1977	11 %	lightning stroke
1978	10 %	" "
1979	2 %	
1980	3 %	
1981	5 %	
1982	1 %	complete sutstitution of data collection units
1983	1 %	

In the time after the lightning strokes the increase of failures as delayed effects was obvious. Together with increasing problems to get substitute components there were reasons enough for a complete substitution of the circuit boards by a micro processor controlled unit. It included also a more power efficient technology (C-MOS instead of TTL), by which failures of the power supply could be reduced considerably.

SOME COMMENTS WITH VIEW TO THE MICRO PROCESSOR CONTROLLED UNIT:

As compared to the "old" system the new system is controlled by a program (soft ware). Thereby a unique flexibility is achieved. For instance: - any of the sensors can be connected or disconnected to the system without interference into the electronic components; - the control of the data flow is improved, which becomes obvious from less errors; - the memory is organized with respect to the sensors, which improves the reliability and flexibility in the read-out of the data. The number of electronic components (hard ware) could be reduced by the use of the microprocessors. The particular circuit boards are modular, which facilitates any modifications or substitutions.

To mention a disadvantage: The great flexibility requires more efforts in the monitoring of the control processes, which may be another source of operating troubles.

COST CONSIDERATIONS

The costs of purchase of the above described data collection and transmission system showed up an amount of approx. $ 35'000 (1982), including 2 field units with 25 sensor connections and one central unit (with possibilities to extend the system). An additional amount of $ 7'500 was needed for the modulation/demodulation device and the automatic telephon dial unit. The maintenance costs for substitutionmaterial amounts to less than $ 500 per year on the average.

The maintenance requirements for staff personnel are as follows:

- The field stations are checked once a week (3 hours)
- The daily processing of the data takes 0.5 hours. Once a month the data are transferred to a permanent tape (ca. 1/2 day/month).

For this maintenance and for repair work one electronic engineer and one technician are available part of their time.

CONCLUDING REMARKS

The operation of an automatic data collection and transmission system over a period of nine years has confirmed its irretrievability to offering a continuous control of the system, including the sensors, and by providing real-time information about of the hydrological events. The great benefit therefore lies obviously in the automatic transmission of the data. At the same time the observation data are available for realtime operational forecasting purposes. The importance of maintenance by trained personnel is to be emphasized. Cost and benefit considerations are clearly on the side of benefit, if manpower is taken into account, which would be necessary without teletransmission to check the recording systems in the field and to collect data necessary for daily operational purposes.

Acknowledgement

Eszter Nyuli provided technical assistance in completing this paper.

REFERENCES

Lang, H. und B. Schädler, 1980: Wasserhaushalt und Wärmehaushalt. Wasser, Energie, Luft - Eau, énergie, air, Heft 9, p. 284-289.

Ott, A., GMBH, Kemten, 1982: Bedienungsanleitung / Montageanleitung (Manual), Anlage Allgomatic.

Büchler, W., 1978: Blitzschutz elektrischer Geräte und Anlagen, Meteolabor.

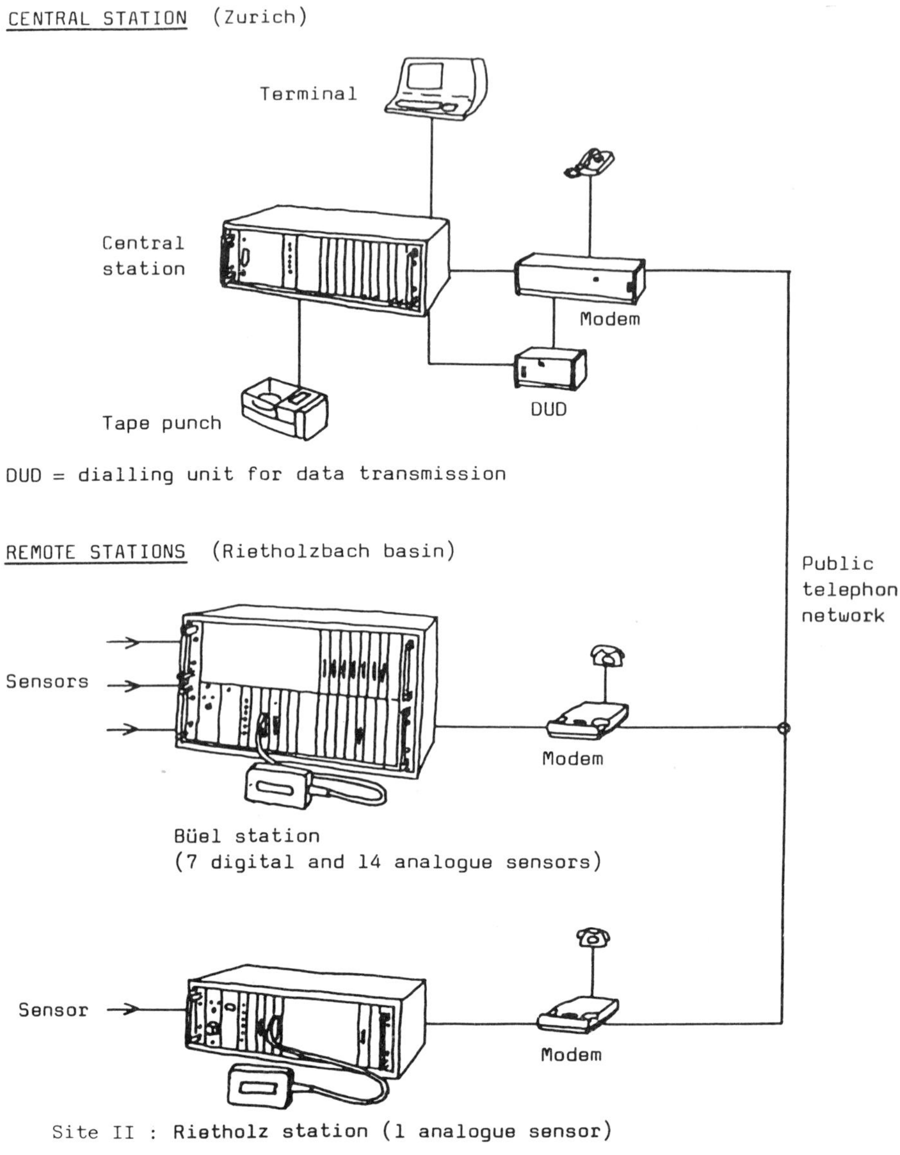

Fig. 1 Automatic data collection and transmission system

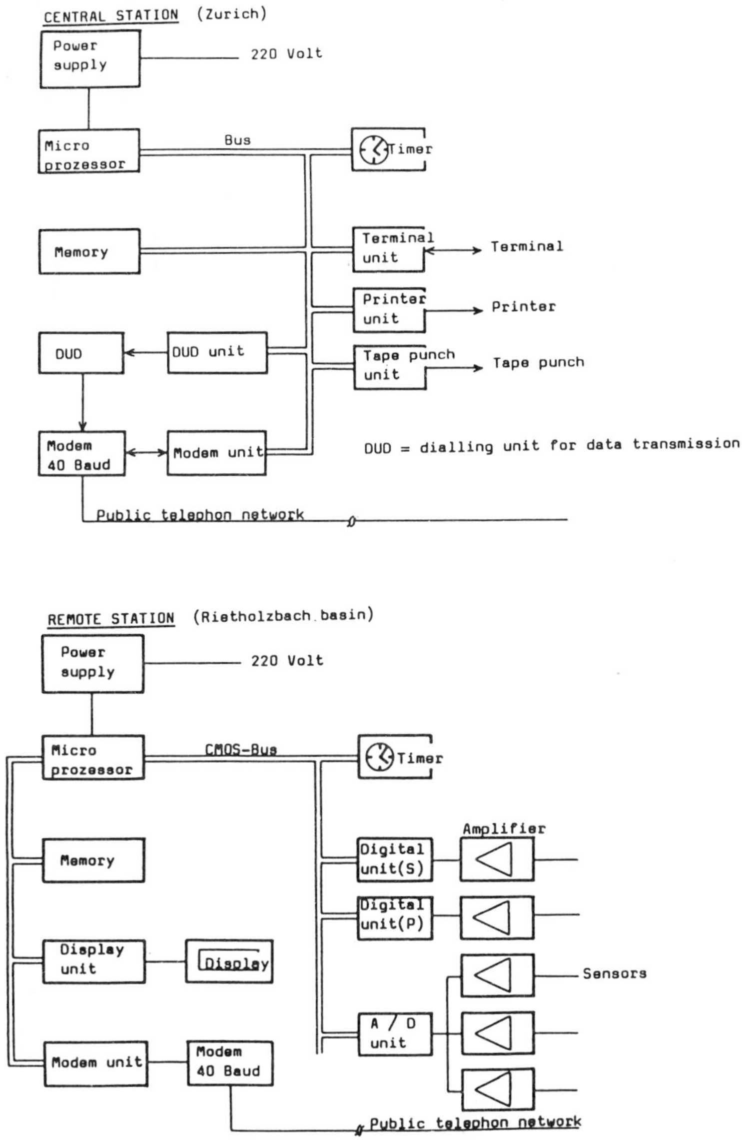

Fig. 2 Block diagram from the central station and the remote station

RIETHOLZBACH-DATEN (ANLAGE ALLGOMATIC) DATUM: 7.MAI 1984

ZEIT	THU	TT	TF	FHU	W2.7	LYG	LYA	NB	ABF
1	90	89	81	956	2	0	1	0	16
2	114	107	91	871	24	0	1	0	16
3	101	94	91	963	49	208	1	70	21
4	98	94	89	963	29	-4	1	0	45
5	97	92	83	960	20	16	1	5	40
6	92	89	78	949	16	14	1	5	28
7	94	91	76	886	11	-4	1	0	25
8	105	99	79	822	10	-4	1	0	24
9	112	108	86	797	18	-2	1	0	22
10	115	110	88	830	23	-4	2	0	20
11	122	113	90	786	7	4	1	2	20
12	129	122	99	774	11	-10	1	0	19
13	123	118	97	812	12	-4	1	0	19
14	127	123	96	774	13	-2	1	0	19
15	133	125	97	796	11	-8	1	1	18
16	116	109	94	890	11	-4	1	0	18
17	93	88	82	960	11	270	1	89	24
18	95	88	85	964	4	78	1	28	90
19	74	67	64	965	8	60	1	21	102
20	55	50	48	963	14	246	1	87	227
21	44	41	39	969	11	60	1	21	417
22	34	26	23	977	16	98	1	33	304
23	25	20	19	970	19	18	1	8	303
24	20	17	18	960	23	0	1	0	229
ZEIT	2400	2400	2400	1400	100	200	2400	2400	210
MINIMUM	20	17	18	774	2	214	1	0	16
ZEIT	1500	1500	1200	2200	300	2300	1000	1700	2010
MAXIMUM	133	125	99	977	49	1240	2	89	484

BODENTEMP. 20CM: 109 / MINIMUM 2400 102 / MAXIMUM 1600 113
BODENTEMP. 40CM: 101 / MINIMUM 1200 100 / MAXIMUM 2400 102

REGENSUMME TOGNINI STANDARD : 35.2 MM
REGENSUMME TOGNINI BODENEBEN: 37.0 MM
SUMME SCHMELZWASSERLYSIMETER: 1 WIPP.
SUMME LYSIMETER AUSFLUSS : 25 WIPP.
SUMME LYSIMETER-GEW. DIFF. : 102.6 KG

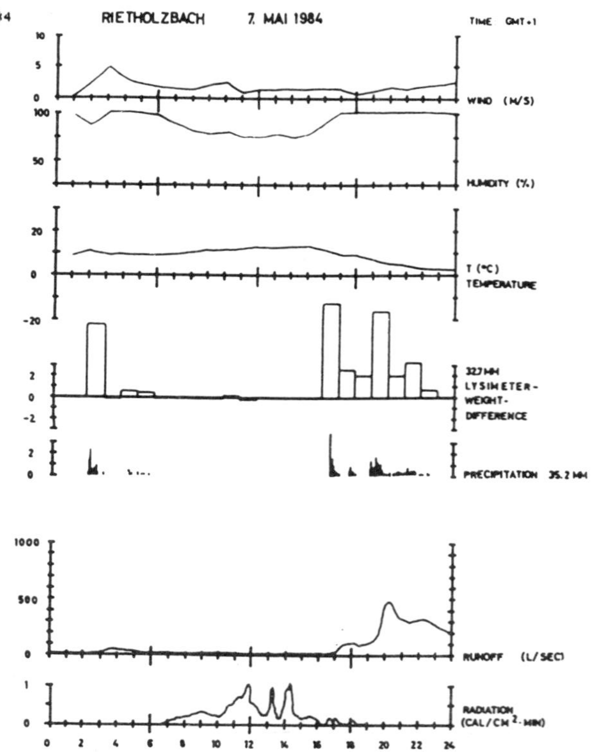

Fig. 3 Example of a daily control output

THU	= Temperature in scree	x0,1°	Celsius
TT	= Psychrometer dry-bulb	x0,1°	Celsius
TF	= Psychrometer wet-bulb	x0,1°	Celsius
FHU	= Humidity in screen	x0,1	per Cent
W2.7	= Wind velocity	x0,1	m/s
LYG	= Lysimeter weight difference	x 0,1	kg
LYA	= Lysimeter discharge	x 50	cm^3
NB	= Precipitation	x 0,1	mm
ABF	= Runoff of Rietholzbach creek	x	L/SEC

COMPUTERIZED SUPERVISION AND CONTROL OF DUJAILAH IRRIGATION PROJECT

Nazar K. Wafi* and Abdul-Ilah Y. Mohammed*

INTRODUCTION

The Dujailah project, located approximately 200 km southeast of Baghdad, Iraq, is an integrated agro-industrial complex. It contains farms and orchards, animal and poultry farms in addition to dairy and food processing industries. The project evolves around a planned central city complete with services including an institute for agricultural technicians.

Its development, planned in three stages, will cover 75 000 hectares. The first stage development which covers approximately 20 000 hectares is now completed. Development of the project included re-construction of the 60 km long main Dujailah canal with intakes on the Tigris river near Kut.

The irrigation network consists of four primary canals totalling 43 km in length and secondary canals 41 km long. Branching from these canals is a network of pipe feeders with hydrants for field plot irrigation. The project employs surface irrigation methods whereby border strips 15 m wide and 284 m long are formed on either side of the pipe feeders. Each hydrant has a basic flow rate of 28 l/s and irrigates a strip on either side (two-sided irrigation). All flow in the system is by gravity. The drainage network follows a similar pattern with covered field drains and secondary and primary open drains. All drainage water flows by gravity to a pumping station at the northeastern corner of the project.

Control of water supply and distribution is accomplished by the Dujailah head regulator (HR) at Kut, four control regulators (CR) on the main Dujailah Canal, 29 regulated and 18 non-regulated (open or closed only) gates on the canal network, Fig. 1. Control Regulator 4 on Dujailah Main Canal controls deliveries to downstream users. The size and complexity of the project preclude manual control and necessitate computerized central supervision and control of the irrigation network. The computerized system is designed for control and supervision as well as management of the water resources of the project. It strives to achieve this through the realization of the operator-system interface and displays.

The Dujailah Supervision and control system consists of the System Control Center (SCC), 52 outstations (one for each control regulator or gate) and the telecommunication network. The basic functions of the system are

- data acquisition
- data and information communication
- processing of data and information
- control, display and supervision of output
- elaboration of operation insturction commands and alarms.

* Professor, Dept. of Electrical Engineering and Asst. Professor, Dept. of Irrigation and Drainage, respectively, College of Engineering, Baghdad University, Baghdad, Iraq.

World Meteorological Organization (ed.), Microprocessors in Operational Hydrology, 213-222.

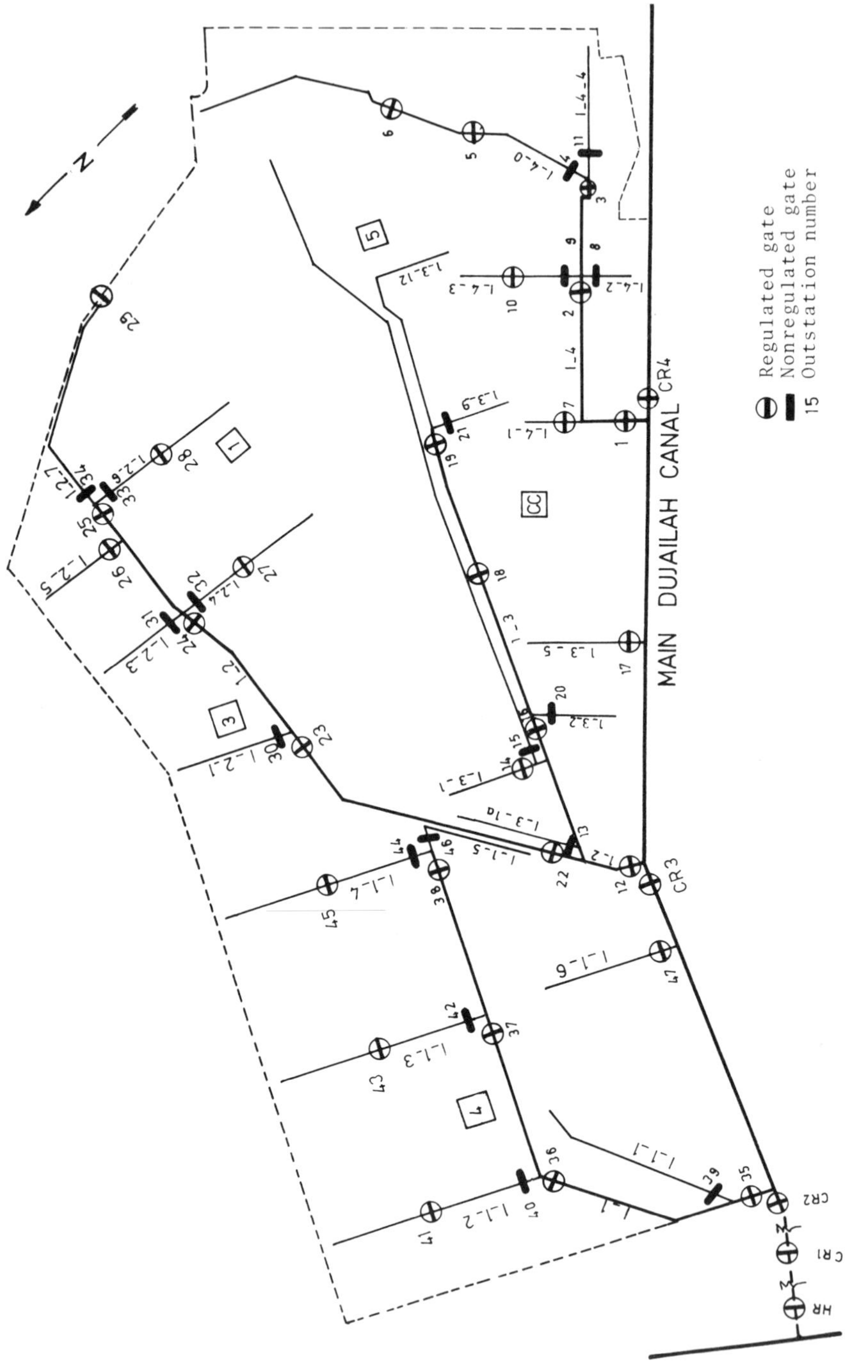

FIG. 1. IRRIGATION NETWORK AND CONTROL STRUCTURES OF DUJAILAH PROJECT

In addition to econo-technical justifications, the design and implementation of the remote control system was subject to the following requirements :

- reliability, data transmission, accuracy and delay restrictions.
- interface with gate and telecommunication network equipment.
- accuracy of transmitted measurements and set point values.
- system flexibility and possibility of future expansion.

The nature of the project area was suitable for downstream type of control whereby water is available upon demand. Automatic operation of the gates system is affected through water level control. Alternatively, it may be accomplished by specifying gate position or discharge. Operation of the gates is system controlled or upon remote commands issued from the System Control Center (SCC). Programs for the simulation of the operation were quite useful at the design phase in testing and finalizing system performance.

THE COMMUNICATION NETWORK

A communication network links the System Control Center with all control regulators and gates on the irrigation canal network and Main Dujailah Canal. Telephone voice communication is also used for point-to-point connection. The information transmitted from System Control Center to outstations include remote commands and analogue set point values. Transmissions from the outstations to SCC include signals and analogue measurements. Signals are indications of the state of the gate and hydraulic structure at the outstation. Details of the communication network's hardware and elaborations of commands, signals and analogue measurements are presented in Table 1.

The slow nature of the hydraulic system to be controlled allows the use of a procedure of cyclic exchanges of information between outstations and the Control Center according to a routine of interrogation-reply modes. This, however, may be momentarily interrupted by priority transmissions. Thus, the system allows mixing of information, e.g., sending supervision and alarm information according to its priority and sending remote control signals and commands. Where information is in transit, it allows a direct connection to a processing unit through use of registers. Hence the operation is based on a continuous dialogue between the computer (and operator) at the System Control Center and outstations at the gates. The transmissions use a binary decimal code (BDC). To eliminate errors due to poor communication channels, the interrogation message addressed to each station consists of two identical messages which are compared to check coincidence. Messages sent by the Control Center include

- synchronization signals enabling the outstation to analyze the message
- the address code of the outstation concerned
- function code

The message sent back from the outstation is made up of

- Synchronization, address and function code identical to that received from the Center
- numerical code containing data measurements and signals

The command and instruction information is achieved in a similar way with the additional information of completing the process and execution.

TABLE 1. DUJAILAH COMMUNICATION NETWORK HARDWARE AND INFORMATION TRANSMISSIONS

SYSTEM CONTROL CENTER	OUTSTATIONS
Hardware Indicators, recorders and displays The signals and alarm unit The remore controllers Interface with computers Command points and instructions	Hardware Remote telemetering unit (RTU) Measurement sensors Position and alarm detectors Automatic control devices and equipment Regulating devices
Number and Types of Transmissions to Outstations Regulated gates 6 Remote commands 1 Analogue set point value Nonregulated gates 4 Remote commands	Number and Types of Transmissions to SCC 4 Analogue measurements 12 Signals (alarms and indications) 10 Signals
Transmission Details	Transmission Details
(a) Remote commands : (1) Start of the gate opening (2) Stop of the gate opening (3) Start of the gate closing (4) Stop of the gate closing (5)*Start of the gate positioning to the previously issued set point value (6)*The start of the water level regulator operation * commands not transmitted to nonregulated gates (b) The analogue set point value is the desired position of the gate.	(c) Analogue measurements : (1) Up-stream water level (2) Down-stream water level (3) The position of the gate (4) The reference value of the local water level regulator. (d) Signals-indications (1) to (6) the corresponding remote command is issued from outstation to the other control equipment at the gate (7) the manual local gate control mode is started (8) the manual local gate control mode is stopped (e) Signals-alarms (9) the gate is at the highest position (10) the gate is at the lowest position (11) main power supply has failed (12) hydraulic failure (low level)

Although most of the network utilizes underground telecommunication cables which run along the power cables, economic considerations dictated exclusion of the Head Regulator at Kut and Control Regulator 1, at km 16.540 of the Main Dujailah, from the underground cable network. Instead, two way double channel UHF radio links are used to transfer information and commands between each of these points and Control Regulator 2, at km 29.654 of Main Dujailah Canal, from which a cable connection to SCC is used.

POWER SUPPLY

The power distribution network consists of underground cables laid along the open channel irrigation network and a major part of the Dujailah Main Canal. It is fed through 11 kV/0.4 kV and 33 kV/0.4 kV substations at eleven points. Increased reliability of power supply is achieved through the use of standby generating sets at the gates to operate the gates. Standby for operating the outstations and communication channels is provided by electronic uninterrupted power supply and batteries. All three types of power supply are used at the Control Center in order to achieve full reliability.

SYSTEM CONTROL CENTER

The basic system control configuration, shown in Fig. 2, satisfies the high degree of computer availability required by on-line applications. It consists of a computer system, a process control console and a programer's terminal for off-line applications. The design of the system and choice of hardware was such that it satisfies current and future requirements. The immediate use of the computer resources is on-line real time supervision and control with limited off-line applications. Future expansion are envisaged to be in control requirements as well as off-line applications.

A dual and reconfigurable computer system is used for on-line real time monitoring and remote control operations. Two 64k words 16 bits core memory mini computers with memory cycle time of 1 μsec are used. Memories are expanded with the addition of 64k words LSI memory to each. Computer CI is a main on-line computer whilst C2 is a standby. Both have the same capabilities and characteristics and are interconnected by an interprocessor link and may exchange on-line data needed for simultaneous updating of data files. Each is individually connected to two 2M words disk memory units, a console terminal (TTY) and two communication modems. The first modem communicates with five outstations on the main Dujailah canal while the second communicates with 47 outstations on the irrigation network. Accommodation of future control requirements of the present and future Dujailah agrocomplex shall affect both CI and C2 computers. This expansion will involve additional modems for communication with future outstations and expansion of the wallboard interface as well as the wallboard itself.

Under normal operation CI is connected to the control console and via modems to the transmission path. It performs the supervisory and data acquisition functions only while C2 is off-line. Failure of C2 does not affect the control system. In case of any detected failure in C1 the maintenance engineer as well as the control operators are alarmed and the following manual reconfiguration procedure commences :

- operation of C1 is stopped, C1 is disconnected from the control

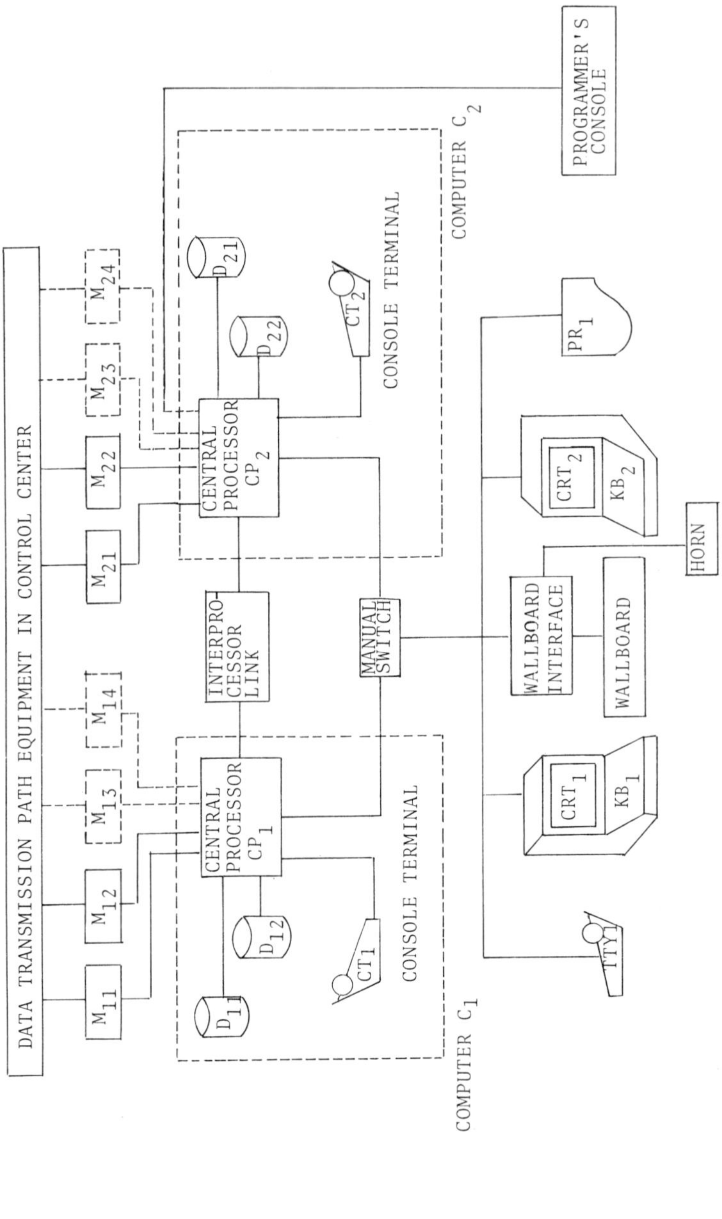

FIG. 2. COMPUTER SYSTEM CONFIGURATION

console and the data transmission path,
- all off-line processing on C2 is interrupted and suspended,
- C2 is connected to the control console and the transmission path and supervisory and control software in C2 is initiated,
- the programmer's terminal is blocked and/or disconnected,
- the C2 takes over control functions.

It is expected that this reconfiguration procedure will last less than 15 minutes. After the failed C1 has been repaired, the reconfiguration procedure is similar : C2 is stopped and outed from on-line system, while C1 takes over on-line functions.

The process control console consists of three semigraphic color CRT displays with associated keyboards, two slow printers (TTY), a fast printer and a mimic wallboard. The console is designed to allow two operators to supervise and control the irrigation system and communicate with other control and maintenance staff in the center or at the gates. It provides a continuous display of the state of the irrigation network. Information is displayed on the CRT's through different images in prespecified format. Two main types of images are provided. The first is an alarm list of present alarms in the entire system. The second is a graphic presentation of a separate gate, or area of canal network, with necessary measured data and alarms related to the displayed part of the system only. The fixed lead line of each CRT image contains the name of the image, the date and time of day. Each image is continuously updated. Normally the alarm list is continuously present on one of the CRT units while the other issues an image only upon the control operator's demand.

Dialogue between the operators and control system, performed through the two functional keyboards, covers issuing remote commands, requests of a specified image on the CRT and request for periodical printouts on the fast printer. The dialogue procedure is simple and does not use a programing language. Hard copy reports of detected alarms and control operators' dialogue through the keyboards are issued on the TTY's. The fast printer issues a datalog report, hourly or upon request, containing all measured values on gates in the system.

The wallboard contains a mimic diagram of the canal network layout. Each gate location is marked with two LED indicators. White LED indicators display the state of the gates (open or closed) while red LED provide failure indications of peripheral and auxiliary equipment. Digital display instruments on the board display the discharge through the primary irrigation network.

The programer's terminal may be connected to Computer C2 only and includes a VDU alphanumeric display with keyboard, a fast printer/plotter, a paper tape reader and a card reader. It is intended for interactive program development and processing which includes off-line programs as well as supervisory and control software.

SUPERVISION AND CONTROL SOFTWARE

The general structure of the system software, which is installed in both computers, is shown in Fig. 3 and may be classified into primary, secondary and special software. The primary software, which includes the primary system executive and the supervisory and control software, may be used only in the computer that is performing the supervisory and control functions (the on-line computer). The secondary software may be used in C2 either when it is a "cold" standby or during the daily break in irrigation (when C2 is off-line). The supervisory and control software consists of four packages : process supervision, process control, recording and reporting, and on-line data files. The first two are project specific and are discussed below.

Process supervision includes :

(a) Process monitoring
- Data aquisition software : control of transmission via modems, transmission analysis, information conversions, storage of information and initialization of programs for information processing, and acceptance of information to be transmitted to the outstations.
- Man-machine communication software : presentation on mimic-board, control of operator-computer dialogue, and displaying on color CRT.
- Primary data processing : basically data filtration which is carried out in continuously shifting time intervals and has characteristics depending on type of noise to be reduced and intended usage
- Logical data checking : utilizes the basic flow equations and certain redundant measurements to determine inconsistencies of measured data.
- Data file storage, updating, and retrieval.

(b) Derived measurements through calculation.
- Water consumption in different parts of the system
- Evaporation losses
- Seepage losses

(c) Logging and alarming in emergency conditions
- Light and sound signals
- Printing of appropriate diagnostic messages.

The general control philosophy adopted for the Dujailah project is downstream control (this does not apply to CR-4 where both upstream/downstream control are invoked since this regulator must satisfy the requirements of downstream users of the main Dujailah canal). The process control software, however, is designed to meet the operational requirements of the system in all contingencies and downstream control may, under prescribed conditions, be overriden and/or supplemented by the following software :

(a) Sequential control. This type of control is envisaged for start-up and shut down. In periods in which large, fast and frequent changes occur downstream control of start-up and shut-down becomes unfavorable. Certain reaches are also very near stability limits and, due to

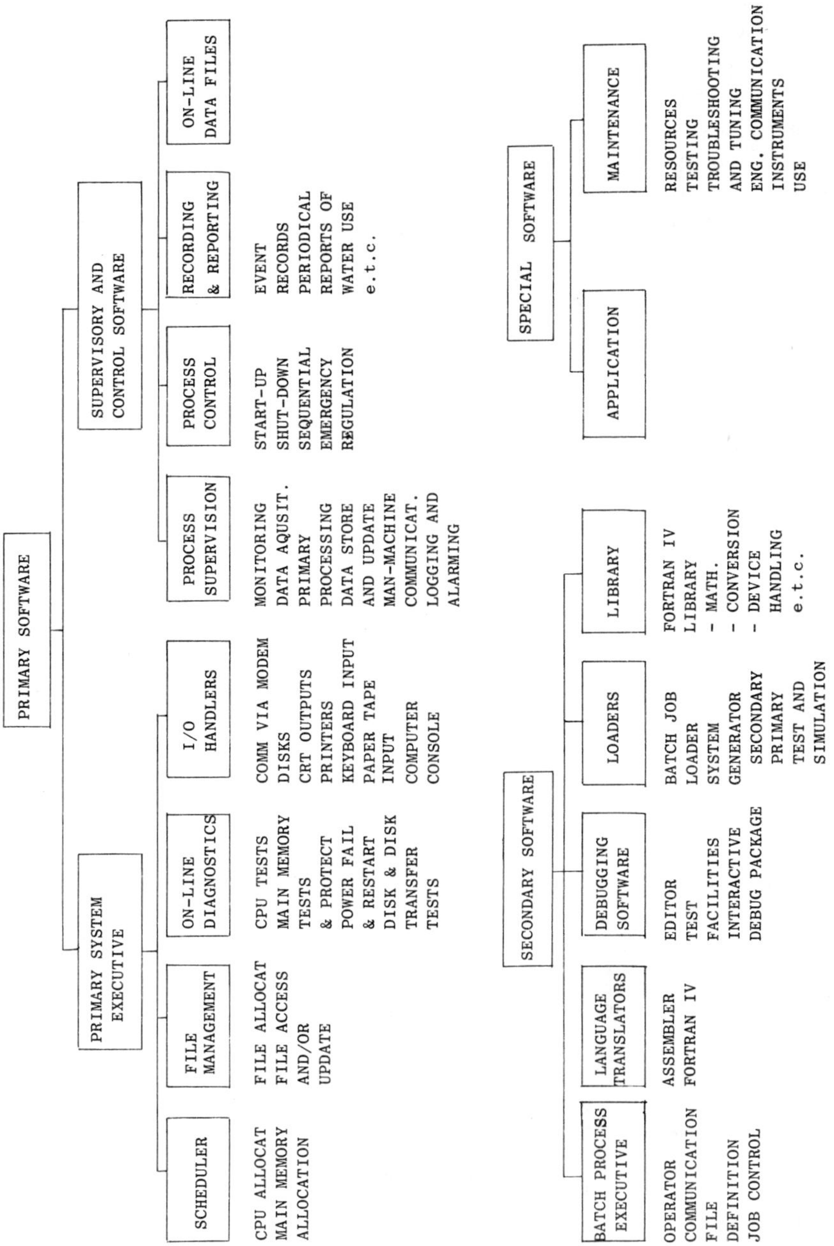

FIG. 3. THE STRUCTURE OF COMPUTER SOFTWARE OF DUJAILAH PROJECT

eventual deviations in designed values, may become unstable. In such cases start-up and shut-down must be realized by sequential control. Further, manual control of the system is easier to realize by a given sequence than by downstream control.

(b) Regulation. In conventional downstream regulation it is most appropriate to have dynamic regulation through the use of on-line real time computer should the need arise. This method not only maintains predetermined water levels at specific locations but may be used to control the rate or volume of flow through each gate. Cases where this software is activated include safety precaution from upstream overflow or when one of the gates is inoperative.

(c) Emergency control. Considering that processes in the canal network occur very slowly, emergency control is non automatic. It is remotely controlled by the operator at SCC and is realized on the basis of corresponding computer messages. Basic emergency conditions envisaged are canal destruction and uncontrolled canal overflow.

(d) Adaptive control. This may come into play in situations where external factors, such as wind velocity or sediment deposition, affect the hydraulic characteristics of the system. No software, however, is presently included.

FAILURE DETECTION

The control system hardware and software allows the detection and indication of certain equipment failures. On-line diagnostic programs run in both computers and the appropriate diagnostic messages are issued on the corresponding computer console. A watch-dog timer detects any abnormality in the operation of computers.

The detection of an abnormal data transmission interrupt between the main computer and any one outstation is performed by a communication software package and indicated via TTY with corresponding lamp indicator on the wallboard and the computer console. Abnormal data transmission interrupt between the main computer and all of the outstations is similarly detected, but the computer system reconfiguration procedure shall be initiated by the control operators.

Certain number of indications concerning the failures of auxiliary equipment in the System Control Center are displayed on a separate panel in the control console. Remotely aquisited data indicating certain abnormal conditions in the gate control equipment are processed in the main computer and the corresponding messages printed via the TTY.

ACKNOWLEGEMENT

The authors are greatfull to His Excellency the Minister of Agriculture and Agrarian Reform for permission to publish this paper. They are indebted to the administration of Dujailah Agrocomplex, owners, and their staff and to Mihailo Pupin Institute of Beograd, designers, who were unselfish and helpfull. Special thanks are due to Mr. K.N. Fattah, chairman of the Irrigation, Drainage and Reclamation Committee of Dujailah, for his commitment to the design team concept and dynamic direction.

THE APPLICATION OF MICROCOMPUTERS TO THE CALCULATION OF DESIGN FLOWS AND FLOOD CONTROL (MANAGEMENT OF SMALL, PARTIALLY URBANIZED CATCHMENTS)

R. Gloor*, P. Wisner**

1. INTRODUCTION

The growing urbanization of rural watersheds has sometimes caused significant changes in the hydrological regimes of rivers. Such changes frequently result in uncontrolled flooding causing serious damage to property and/or human injury.

Other factors, however, such as the non-regulated spread of plastic greenhouses in rural areas or the construction of major highways, can have similar effects on hydrology.

Many examples of this phenomenon may be found worldwide. The case of the Seymaz (located on the outskirts of the city of Geneva, Switzerland) is described in this paper because the site is familiar to the authors. The urbanized region lies downstream of a very flat watershed (1.9%). The current urbanization rate is 14% and is likely to increase to 25% within the next 20 years. Serious flooding occurs somewhat frequently.

When hydrological and hydraulic studies are undertaken on such watersheds, complex problems rapidly arise which can only be resolved through computerized data processing and simple or complex hydrological models. Complex models, by their very nature, are generally developed and accessible on mainframe computers. Consequently, they are only available to specialists with access to mainframe equipment. These sophisticated calculatation methods can only be made available to all users if they are adapted to microprocessors.

Actually, in the field of hydrology, microprocessors can and should serve at two levels:

1. The engineering firm, during the preparation of preliminary studies and detailed projects;
2. The administrative level, when supervisory activities are required for complex projects, for the management and analysis of planning stages, and even for real-time forecasting.

Although engineering firms generally have in-depth knowledge of hydrology, such is not the case at the administrative level. A model made available to the "supervisor", therefore, must be simplified and must produce results which are compatible with the methods used by the "designer", and must be easily accessible. Clarity and consistency are, therefore, essential elements.

* Lecturer, Institute of Rural Engineering, Ecole polytechnique fédérale de Lausanne, 1024 Ecublens VD, Switzerland.
** Professor, Department of Civil Engineering, University of Ottawa, 770 King Edward Avenue, Ottawa, Ontario, Canada.

World Meteorological Organization (ed.), Microprocessors in Operational Hydrology, 223-232.

2. THE OTTHYMO MODEL

Among the vast numbers of hydrological models available on the market today, the OTTHYMO model (Ottawa Hydrologic Model, Jordan J.-P. et al. 1983) was chosen for our study. The OTTHYMO model is an expanded version of the HYMO (Williams and Hann, 1973), and includes four new subroutines:

- URBHYD for urbanized areas;
- NASHYD for rural areas;
- KINROUTE for calculating flow in collectors;
- DUHYD enables runoff on roads and in collectors to be considered separately in urban areas.

The OTTHYMO model may be used for the preparation of overall plans or for detailed studies.

For example, the URBHYD subroutine (P'ng, 1982), may be used for the respective calculation of runoff from impermeable and permeable parts of an urbanized catchment. This would be a conceptual model with two parallel linear reservoirs. URBHYD determines two hyetographs of excess rainfall from any hyetograph of input rainfall. With regard to impermeable surfaces, the only losses to be taken into consideration are losses in depressions. The excess rainfall hydrograph of permeable surfaces is obtained after infiltration losses have been deducted. Infiltration losses may be evaluated either by the Horton equation, or by using the CN* procedure, which is a modified version of the method developed by the U.S. Soil Conservation Service (SCS) in 1971. The two unit hydrographs may then be calculated from the storage coefficients of the two linear reservoirs. The unit hydrographs are then combined with the corresponding excess rainfall hyetographs in order to obtain the respective hydrographs. The two α and β, scale factors, which depend on the impermeable and permeable areas respectively, multiply the two unit hydrographs and once the two hydrographs are added together they form the resulting hydrograph of the whole subwatershed.

Thus, one of the primary advantages of the URBHYD lies in the ability to take into account an increment in impermeable surfaces by increasing coefficient α.

The storage coefficient used in URBHYD is calculated from an equation used by Pederson et al. (1980) which takes 50% of the equilibrium time datermined by the kinematic wave theory. The storage coefficients are peculiar to each linear reservoir and thus take into account the differences in time of concentration between the impermeable and permeable surfaces.

The OTTHYMO model has been tested on three urban watersheds and the results have been compared to measurements taken.

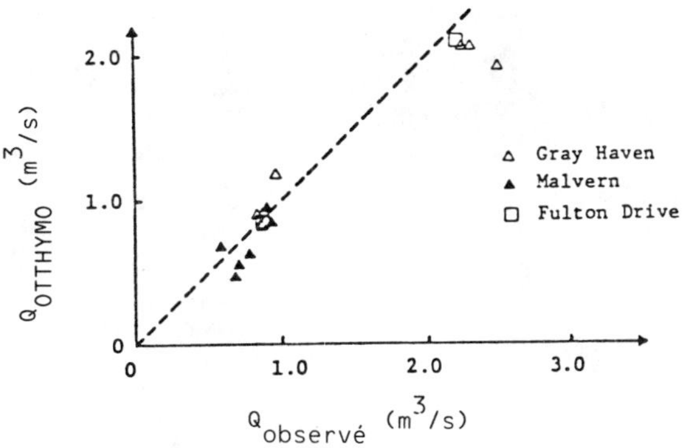

Figure 1 - Comparison of calculated and measured peak discharges

In areas where the SCS methods's CNs are low, the initial losses are generally overestimated by the SCS, and the calculated surface runoff of small rain events is often insufficient or non-existant. The NASHYD sub-routine, which is a linear model, uses a modified version - the CN* procedure (Jobin, D., 1982). The initial loss is thus obtained as a function of the Antecedent Precipitation Index (API). This procedure, which was developed and applied to the Seymaz watershed, has been successfully tested in five rural catchments in Canada. Moreover, it has been shown in rural catchments that the parameter n varies between 2.3 and 3, so that only peak time t_p need be calibrated.

3. AN APPLICATION OF THE OTTHYMO MODEL

The OTTHYMO model has been applied to the Seymaz basin in a joint study undertaken by the Institute of Rural Engineering of the Ecole polytechnique fédérale de Lausanne and the University of Ottawa (Gloor, R. et al., 1983). The Seymaz basin has a catchment area of 38.4 km^2, of which 8.2 km^2 are in an urban area at the downstream end.

Five recording raingauges and four water-level recorders with a 10-minute resolution have been in place on the Seymaz watershed since 1976. Detailed measurements were therefore obtained from a recent flood, with a return period of approximately 25 years. The photograph in Figure 2 demonstrates the extent of the flooding.

Figure 2 - Seymaz flood, Geneva, January 1979

Modelling of this watershed was undertaken in two stages:

- The rural zone was modelled first for which the calibration was facilitated by the excellent measurements taken at Pont Bochet which is the dividing line between the rural and urban areas. It was therefore possible to determine the peak time, the initial losses and the CN*-API curve (Figure 3).

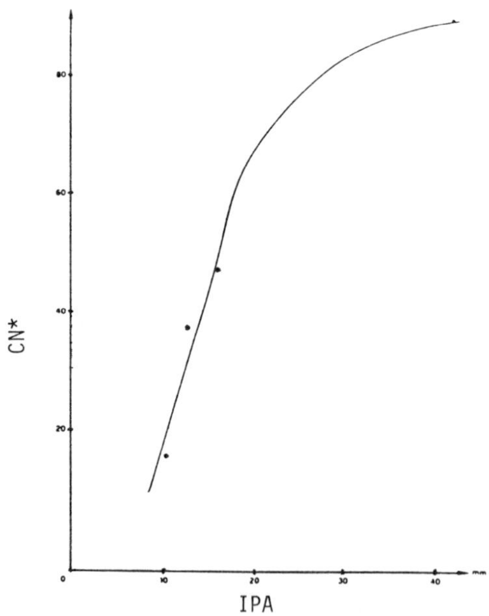

Figure 3 - CN*-API ratio

- The outlet measurements (Figure 4) were then used to model and validate the entire watershed.

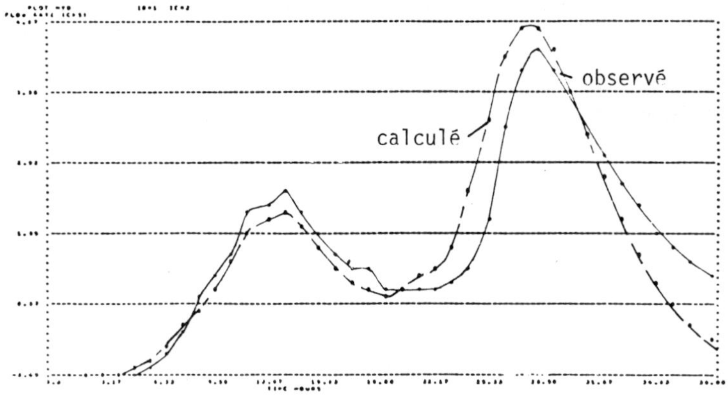

Figure 4. Comparison between the calculated and the observed Hydrograph. (Villette 27 May 1979)

The modelling of historic rain series associated with the corresponding initial losses makes it possible to establish a flood frequency curve. The simulation of short- and long-term critical rains was useful in the comparison of alternatives. The calibrated model also served to evaluate the influence of an impermeability increase in the urban zone and to compare alternatives - canalizations, embankments, upstream retentions, diversion works, etc.

This study was undertaken on a mainframe computer and the results were compared with the OTTHYMO microcomputer version described below.

4. THE OTTHYMO MODEL "MICROCOMPUTER" VERSION

The first PC-PTTHYMO (Personal Computer OTTHYMO) version was written in FORTRAN and functions on an Apple II. It was designed for planning departments which already have some experience in modelling procedures. During one Seymaz comparison study, hydrographs calculated by the PC-OTTHYMO were practically identical to those previously obtained from the OTTHYMO method. During the OTTHYMO study, the urban zone had been subdivided into 18 subwatersheds whereas during the PC-OTTHYMO study, the zone was reorganized into eight subwatersheds. An example of the results obtained is shown in Table 1:

	Number of subwatersheds	Peak discharge (m^3/s)	Peak time (h)
OTTHYMO	18	12.49	1.7
PC-OTTHYMO	8	12.27	1.7

Table 1 - Comparison between OTTHYMO and PC-OTTHYMO

During this comparison study, the NASHYD subroutine was also successfully employed for modelling the rural zone.

Based on this experience, an interactive PC-OTTHYMO version was prepared and Figure 5 below demonstrates the simplicity of data input on the NASHYD subroutine.

```
                    NASHYD SUBROUTINE - 1.1
                    -----------------------
FOR DEFAULT VALUES OF (CN), (N) AND (IA) THEN HIT "RETURN"
<CR> = HIT "RETURN" AFTER INPUT
--------------------------------------------------------------------------

Please ENTER the following:

>1) UNITS, metric or imperial  (M,I)            <CR> =
 2) CURVE number                                 <CR> =
 3) AREA of watershed                            <CR> =
 4) Number of NASH RESERVOIR   (DEFAULT=3)       <CR> =
 5) Time to PEAK (hr)                            <CR> =
 6) Initial ABSTRACTION                          <CR> =
 7) NAME of STORM to be used                     <CR> =
 8) OUTPUT on line PRINTER     (Y,N)             <CR> =
 9) Print UNIT HYDROGRAPH ordinates  (Y,N)       <CR> =
10) Print RAINFALL EXCESS      (Y,N)             <CR> =
11) Print FINAL HYDROGRAPH     (Y,N)             <CR> =
12) Enter NAME for OUTPUT FILE (DEFAULT=HYD1)    <CR> =
```

Figure 5 - Data input (NASHYD subroutine)

Some recent PC-OTTHYMO applications are summarized in Table 2. The model was sufficiently tested and the results were compared to results from mainframe models. Small engineering firms not equipped with mainframe or minicomputers derive numerous advantages from the PC-OTTHYMO method, such as simple editing and data storage procedures, and rapid print-outs. Clients can thus receive reports at the same time as basic data and the adapted model, which is recorded into a diskette, thus facilitating re-use if changes are necessary. Moreover, microcompuuters can be used for other functions, such as word processing, invoicing, etc., and hardware costs may thus be distributed among several departments. However, certain limits are imposed by the computation and printing time - see Table 2 for orders of magnitude. In order to reduce computation time in favour of an analysis of several project rains or alternatives, the number of subwatersheds should be limited to 15. In some cases a less detailed discretisation will be necessary.

As mentioned above, the model offers the possibility of parameter calibration, such as the CN curve number, the number n of reservoirs, peak time and initial loss. Assuming, however, that all these elements have been previously defined and that the project rains were chosen (real or synthetic), a simplified version can be made available to an administrator, thus enabling him to undertake simulations by using the planning parameters only (surface, impermeability or channel flow).

In fact, for any given region, it is possible to determine which of the parameters in a model are approximately constant in time. It is also possible to choose project rains serving as a basis for NASHYD or URBHYD. In such cases, a very simplified OTTHYMO version (known as SIMPHYD (Wisner and Lam, 1984)) may be used to calculate a hydrograph corresponding to an impermeabilization rate and a given CN* value, i.e. a determined planning state.

Job Name	Total area studies (ha)	Type of study	No. of subwatersheds	% of urbanization	No. of online storage	Duration of computation for design condition
1. Moreau Creek, Gatineau	1,019	MDP	13	50	2	1 hr.
2. Morningside, Markham	1,012	MDP	9	90	3	3/4 hrs.
3. Hotchkiss Creek, Barrie	412	MDP + FC	14	100	1	1 1/2 hrs.
4. Lover's Creek, Barrie	6,100	MDP + FC	15	100	3	2 1/2 hrs.
5. Dyment Creek, Barrie	532	MDP + FC	11	80	1	1 hr.
6. Brand Development, Peel	30	MDP	5	100	1	1/4 hrs.
7. Vaughan 400, Toronto	796	MDP	12	46	2	3/4 hrs.
8. Rouge Industrial	156	MDP	12	80	2	3/4 hrs.
9. East Hager Creek	96	MDP	5	100	1	1/4 hrs.

MDP Master Drainage Plan
FC Flood Control Study

Table 2 - Examples of studies undertaken on microcomputers

Managers or administrations without much knowledge of modelling procedures may use this simplified model to undertake simulations. The model has been successfully tested against the PC-OTTHYMO and the OTTHYMO on numerous occasions. An added advantage is the vastly simplified input and output procedures (Figures 6 and 7).

The case of Morningside, Markham (Toronto) may be cited as an example. Use of the PC-OTTHYMO made it possible to calibrate the various parameters and to check them. It was then possible to integrate these parameters into a simplified version of the model, thus enabling the municipality to undertake check calculations and to analyze new situations.

```
                MICROCOMPUTER APPLICATIONS
                 IN URBAN WATER RESOURCES
               COORDINATOR: DR PAUL WISNER
                 UNIVERSITY OF OTTAWA.

                           BY:
             DR P.WISNER, A.LAM, J.F.SABOURIN

          * PRELIM. COPY (PILOT PROJECT)
          * TOWN OF SCARBOROUGH

---------------------------------------------------------------------------

DESIGN STORM USED = YCH.STM

                    RAINFALL DISTRIBUTION (MM/HR)

             8.808   10.614   16.421  161.694   29.676
            18.763   14.121   11.487    9.768    8.547
             7.631    6.915

NOTE: <*> = Partial hydrograph at junction before adding the
            subarea or tributary flows.

                 SUMMARY OUTPUT FOR JONCTION POINT NO. 1
                 ---------------------------------------

WATERSHED    AREA    IMP    CN    S      PEAK    TPEAK    RUNOFF VOL
              ha.     %           %      cms      hrs        mm
  R.a.     200.00    **     65   1.0     4.77     1.50      13.46
  R.b.     100.00    **     65   1.0     3.15     1.17      13.47
======    ========                     =======  =======   ========

  JN 1     300.00    **     **    **     7.50     1.33      13.46
CHANN-ROUTING TO:   ( Convex Method: L=1000 m, v=.5 m/s )
  JN 2     300.00    **     **    **     6.43     1.83      13.26

                 SUMMARY OUTPUT FOR JONCTION POINT NO. 2
                 ---------------------------------------

WATERSHED    AREA    IMP    CN    S      PEAK    TPEAK    RUNOFF VOL
              ha.     %           %      cms      hrs        mm
  R.a.     200.00    **     65   1.0     4.77     1.50      13.46
  R.b.     100.00    **     65   1.0     3.15     1.17      13.47
  <*>      300.00    **     **    **     6.43     1.83      13.26
======    ========                     =======  =======   ========

  JN 2     600.00    **     **    **    12.63     1.67      13.36
```

Figure 6. Example of PC-OTTHYMO output

```
HYDROGRAPH FROM : JN
               TOTAL AREA =   600.00 HA
               PEAK FLOW  =    12.63 CMS
               TM TO PEAK =     1.67 HRS
               RUNOFF VOL.=    13.36 MM

TIME:0.000   0.167   0.333   0.500   0.667   0.834   1.000   1.167
FLOW: 0.00   0.00    0.01    0.05    1.11    3.86    6.91    9.57
-------------------------------------------------------------------
TIME:1.334   1.500   1.667   1.834   2.000   2.167   2.334   2.500
FLOW:11.46   12.44   12.63   12.22   11.44   10.41   9.13    7.75
-------------------------------------------------------------------
TIME:2.667   2.834   3.001   3.167   3.334   3.501   3.667   3.834
FLOW: 6.36   5.06    3.92    2.96    2.18    1.58    1.13    0.79
-------------------------------------------------------------------
TIME:4.001   4.167   4.334   4.501   4.668   4.834   5.001   5.168
FLOW: 0.55   0.05    0.03    0.02    0.01    0.01    0.00    0.00
-------------------------------------------------------------------
```

Figure 7. Hydrograph calculated by PC-OTTHYMO

5. FINAL NOTE

It should be noted that, in the case of mixed subwatersheds, two or more types of characteristic and critical rain events should be taken into account, for example:

- Short, but every intense, rain events;
- Very long, but less intense, rain events;
- Etc.

Their respective duration corresponds approximately to the time of concentration of one part of the watershed or another, i.e. the rural or urban zone.

But even if the separation of these two types of rain is very important when estimating peak discharge, the preparation of their project hyetographs remains very complex, unless strict hypotheses are formulated.

6. CONCLUSIONS

The models described in this paper for use on microcomputers are an invaluable tool. They function interactively and are very easy to use, both at the data input level as well as at the result analysis stage. Moreover, they have been used and verified on numerous occasions in Switzerland and Canada. These models are available to:

1. Project engineers, who generally do not have access to complex models, (either due to little hydrological knowledge or a lack of available data processing facilities - except in the case of very large firms), and are therefore frequently reduced to using simplistic methods, such as the rational formula;

2. Supervisory administrations, who must be in a position to rapidly and accurately check and evaluate the projects submitted to them, or to analyse the impact of new constraints (e.g. changes in impermeability rates);

3. The models can also be used to evaluate a situation in real time.

In all cases, it is essential that the calculation methods employed are easy to implement and are perfectly clear. The user must be able to understand what he is doing. Furthermore, it is essential that consistant and compatible results be obtained.

The PC-OTTHYMO and SIMPHYD models appear to fulfill these requirements.

7. ACKNOWLEDGEMENTS

The authors wish to thank Mr Alan Lam, who helped adapt these programmes to microcomputers; Mr J. Sabourin, who facilitated access to the programmes; and Mr Andrew Brodie, of Andrew Brodie Assoc. Inc., Toronto, who actively participated in the developments.

REFERENCES

Gloor, R. et al., Etude Hydrologique du Bassin Versant de la Seymaz (Hydrological study of the Seymaz watershed), IGR 173, Ecole Polytechnique Fédérale de Lausanne et Université d'Ottawa, 1983.

Jobin, D.I., Application du Modèle OTTHYMO sur un Bassin Expérimental Suisse (Application of the OTTHYMO model to an experimental Swiss catchment), thesis, M. Eng., Department of Civil Engineering, University of Ottawa, 1982.

Jordan, J.P. and Wisner, P., Description du modèle OTTHYMO et exemples d'application (Description of the OTTHYMO model and applications), IGR No. 172, Ecole Polytechnique Fédérale de Lausanne et Université d'Ottawa, 1983.

Pedersen, J.T., Peters, J.C., and Helweg, O.J., Hydrographs by Apple Linear Reservoir Model, Journal of the Hydraulics Division, ASCE, Vol. 106, No H5, Proc. Paper 15430, May 1980, pp. 837-852.

P'ng, C.E., Conceptual Hydrologic Modelling for Master Plans in Urban Drainage, M.A.Sc thesis, Dept. of Civil Engineering, University of Ottawa, unpublished.

Soil Conservation Service, National Engineering Handbook, Section 4, Hydrology, U.S. Department of Agriculture, U.S. Government Printing Office, Washington, D.C., 1971.

Williams, J.R., and Hann, R.W., HUMO, A Problem-Oriented Comupter Language for Hydrologic Modelling, User's Manual, ARS-S-9, U.S. Dept. of Agriculture, May 1973.

Wisner, P.E., A Hierarchical Approach to Urban Hydrologic Modelling: The IMPSWM Methodology, 1983 International Symposium on Urban Hydrology, Hydraulics and Sediment Control, University of Kentucky, Lexington, Kentucky, July 1983.

Wisner, P.E. et al. IMPSWM Urban Drainage Modelling Procedures, 2nd edition, The Implementation of Storm Water Management Program, University of Ottawa, 1983.

Wisner, P.E., Fraser, H., and P'ng, C., Comparison of OTTHYMO and VUH-HYMO, Report to Ministry of Natural Resources, Department of Civil Engineering, University of Ottawa, Ottawa, 1983.

Wisner, P.E., and Lam, A., Twin Models for Storm Water Management on Mainframe Computers and Microcomputes, ASCE Engineering Foundation Conference on Emerging Computer Techniques in Stormwater and Flood Management, Niagara-on-the-Lake, Ontario, Canada, October 30 - November 4, 1983.

Wisner, P., and Lam, A., First Short Notes on Microcomputer Application on Urban Hydrology, Andrew Brodie Associates Inc., Ottawa, 1984.

Wisner, P.E., Lam A., and Chin, N., DUHYD - Development of Dual Drainage Submodel in OTTHYMO, Internal Report, Department of Civil Engineering, University of Ottawa, 1983.

Wisner, P.E., and P'ng, C., OTTHYMO - A Planning Model for Master Drainage Plans in Urban Areas, Proceedings, First International Seminar on Urban Drainage Sytems, Southampton, UK, Pitman Advanced Publishing Program, 1982.

HYDROLOGY AND DATA PROCESSING IN THE MEMBER STATES OF THE COMITE INTERAFRICAIN D'ETUDES HYDRAULIQUES (CIEH)

by C. Puech*

The CIEH (based in Ouagadougou, Burkina Faso) is situated roughly at the geographical centre of its 13 Member States. It serves as a focal point for hydraulic studies in the various countries, either by centralizing and redistributing hydraulic information (the documentation centre has 12,000 documents), or by carrying out or commissioning studies on hydraulics which are of interest to all or some of its Member States.

1. INTRODUCTION OF COMPUTERIZED DATA PROCESSING WITHIN THE CIEH AND ITS MEMBER STATES

The CIEH carries out some of the studies included in its programme itself, and in respect of hydrology alone it has had to computerize its data processing for this work. Various programs have been developed on microcomputers (in BASIC) for the following applications:

- Statistics, data analysis
- Fitting statistical distributions used in hydrology
- Mapping, automatic plotting of stage curves (see example in Figure 1)
- Runoff models
- Etc.

These applications have been developed for specific studies such as the study of exceptional (24-hour) rainfalls in the various states, development of a new method for determining flood discharges in small basins in the region, automatic mapping of control measurements for the artificial rain experiment held in Ouagadougou in 1983, calculations of areal reduction factors for rainfall and calculations of urban runoff, secondary processing of rainfall records in order to plot "rainfall depth-duration-frequency" curves (see example in Figure 2), etc., all of these applications being of a very distinctly scientific nature.

Moreover, because of its role as advisor to the Member States, the CIEH is concerned with the computerization of the data-processing services in the relatively new African States. It helps them chose the computer hardware or in developing software, such as programs for publishing the hydrological yearbooks for the Ministry of Rural Development of Burkina Faso. In collaboration with the Ecole Inter-Etats d'Ingénieurs et de l'Equipement Rural (EIER, Ouagadougou), it has also held training courses on data processing to help the Member States master the relevant techniques.

2. DIFFICULTIES PECULIAR TO AFRICA

The CIEH has thus acquired a certain amount of experience through its own activities and its contacts with the States in data-processing applied to hydrology in the African context. It should perhaps be emphasized that this concerns the storage and processing of data and not the field of measurement equipment.

* Chief of the Hydrology Department of the CIEH

World Meteorological Organization (ed.), Microprocessors in Operational Hydrology, 233-237.

It must be said that, apart from two or three capitals with representations of serious firms, the rest of west and central Africa is at present approaching microprocessing in an anarchic fashion. There are a few rare cases of centralized systems which take many years to become operational (too many, given the rapid development of data-processing systems), but largely the Services are acquiring different types of hardware with no possibilities of local maintenance. Indeed, frequently, the Service has received a "black box" system from abroad which it is barely able to master and of whose possibilities of transformation or evolution it is totally unaware.

In Ouagadougou, the problem has been solved in a different way. Because of the impossibility of local maintenance for data-processing hardware owing to the total lack of firms' representatives, it has been agreed jointly that the local bodies concerned with hydraulics would endeavour to acquire the same type of microcomputer, as far as possible. Mutual assistance with regard to both hardware and software will obviously facilitate for the proper operation of the system.

It should also be said that the local climatic conditions are scarcely favorable to sophisticated equipment. Excessive temperatures and dust in the Sahel damage the magnetic media (diskettes) more quickly than elsewhere, humidity attacks the hardware in tropical or equatorial regions, and there is the problem of the stability of the electricity supply, whose voltage may vary from one moment to the next to an astonishing extent (40 to 260 V for the normal of voltage of 220 V). Without wishing it, these countries are thus actually a testing bench for the hardware, which is really proving to be quite robust.

Different constraints from those in Europe may play a part in the computerization of data processing. They are mainly human constraints due to the fact that the Services themselves have very few technicians capable of operating this equipment, and the few commercial representations place the main emphasis on sales rather than maintenance (for the moment), and the choice of equipment in Africa is therefore very problematic.

3. PROBLEMS SPECIFIC TO HYDROLOGY

In our opinion, the most acute current problems are the following:

(1) Difficulty of acquiring software for hydrological purposes (cost, slowness of delivery and performance), and the cost or, in other words, the development time needed to achieve even modest performances if this software is developed by the Services themselves. The number of usable programs is therefore very limited. In this context, WMO's activities under the HOMS programme is therefore very praiseworthy, since it permits great savings by offering efficient programs at moderate cost. The CIEH has in fact offered to become the HOMS focal point in the subregion.

However, it must be fully understood that only programs which are simple to use and very general in application can be really useful. A certain number of programs may be too specific to a given Service or region and difficult to transfer elsewhere. Contrariwise, others are simply a translation of a simple formula and it is difficult to see the need to seek them out.

(2) No less important than the above problem is that of the exchange of information between computers. The difficulties can be broken down as follows:

- Problems of physical compatibility between hardware or between media (which are difficult to solve with 5" diskettes). In the case of 8" diskettes the IBM format seems to be becoming standard. Let us hope that this becomes generalized.

- Problems of language compatibility, in particular for BASIC and its many "dialects" which cause a considerable waste of time in adapting the programs from one type of computer to another and even from one range to another within the same make.

- Problems of homogeneity of data format.

We shall now look at an example of the problems caused by the computerization of data processing in the Services. In the case of Hydrological Services, there are at present very few efficient programs offered to the States. Those wishing to use the possibilities of computer processing therefore have to try to solve the problems themselves for their own particular case. This is the case, for example, with the production of annual reports, processing of streamflow gaugings, etc. As a result, there is a multiplicity of programs for apparently similar basic purposes but which prove in fact to be quite different. This software is generally not compatible either from the point of view of the hardware used or from that of its design, which leads to the storage of partially incompatible information.

For example, in the Ivory Coast, the Surface Water Resources Division works in BASIC on a WANG minicomputer and stores its stage data using variable time steps. On the other hand, Burkina Faso works in BASIC on an Apple III microcomputer and stores its stage data using a fixed time step, taking up to six values per day. Mali, on the other hand, works in BASIC on the PDP 11/34 computer belonging to the AGRHYMET Programme but stores its information using a fixed time step of only one value per day. There are thus obvious difficulties in collecting all of this information in a homogeneous form, unless the least efficient version is taken, but with a certain loss of information from the other countries. This situation results from the historical practices in the various States as regards storage of information. Formerly, these operations were carried out by the local ORSTOM centres with more or less sophisticated FORTRAN programs, depending on the particular State. Even if links between the computers are set up, one can easily imagine the problems involved in centralizing these data in a coherent form.

(3) Problems in the choice of hardware for storing information. The solution of 5" or 8" diskettes which is proposed for commercial microcomputers is generally insufficient for the large measurement networks of the African countries because of both the volume (too much information and therefore too many diskettes) and the storage security (too rapid deterioration of the diskettes).

4. ACTION TO BE UNDERTAKEN

This meeting is an opportunity for us to say how much we favour concerted action between the different States, to attempt to homogenize the hardware and software. This will help save time and effort but will also make it possible to use programs which have been developed by specialists and are therefore, in principle, more efficient than those which the national Services could develop themselves.

The system set up by AGRHYMET, although not yet really operational, is interesting in this respect, since it covers the eight countries of the CILSS (Comité Interafricain pour la lutte Contre la Sécheresse au Sahel); the CIEH would be prepared to undertake any similar action to complement that of the CILSS countries.

Moreover, we can only welcome the HOMS Programme, especially since the cost of data-processing hardware is now becoming relatively insignificant compared with that of developing operational, efficient software.

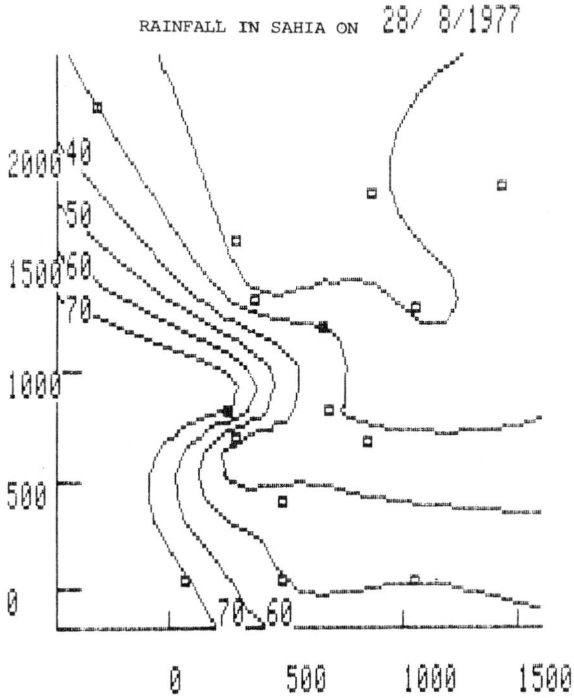

Figure 1
Example of automatic plotting of isohyets

Figure 2

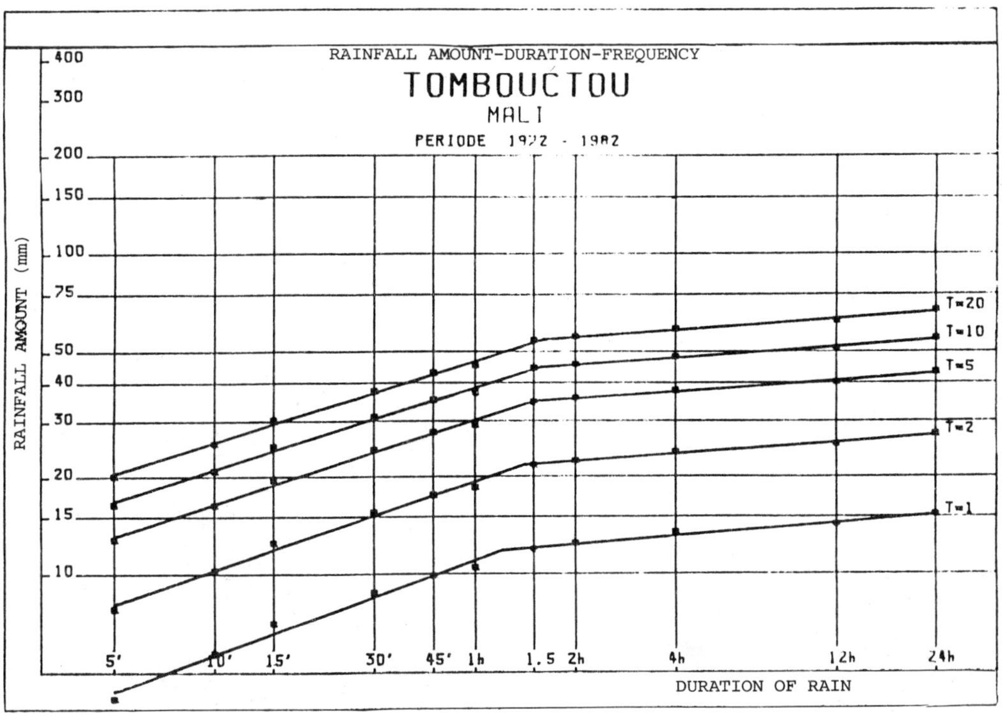

COMPUTATION OF DESIGN FLOOD LEVELS ON MICROCOMPUTER

S. Bognár* and L. Somlyódy**

INTRODUCTION

One of the most important information for planning flood control projects is the design flood level which is generally determined either on the basis of observations or by calculating the steady water surface profile. The flood levels are, however, rarely associated with steady flow since at the time of flood peak the water level is also affected by the variation of discharge. When flood control measures - like the design of flood dykes, diversions, reservoirs etc. - are under consideration their effect on flood level should be predicted, otherwise proper sizes /e.g. elevation of dyke crests/ can not be determined with satisfactory accuracy.

Unsteady flow computations are, therefore, necessary to derive design flood level and changes in flood levels. There are number of methods to solve the unsteady flow equations /Chow 1959, Mahmood and Yevyevich 1975, Kozák 1977 etc./. Most of the solutions had a common drawback: they needed a relatively large computer core memory, and considerable execution time. Their application for design flood calculations was, thus, hindered especially for small scale projects.

By the introduction of micro /and personal/ computers the application of the unsteady flow calculation becomes more economical, but software changes are necessary due to limited RAM capacity of microcomputers.

In the present case a small HP-85 microcomputer of 32 kByte RAM was available, thus necessitating to employ an effective solution procedure for the Saint-Venant equations.

The requirements of the calculations were as follows:

- the irregularity of the natural channel had to be taken into account;
- the boundary conditions of the model had to be consistent with the available geometric and hydrologic data;
- the procedure had to lead to an acceptable execution time on the microcomputer employed.

Instead of the traditional methods for solving the system of linear equations resulted in from the use of the implicit finite difference method, the computationally effective "double-sweep" method was applied. Data required for the computations were stored on mini discs.

* Research associate ** Deputy Director

Institute for Hydraulic Engineering, Research Centre for Water Resources Development VITUKI, Budapest, Hungary.

World Meteorological Organization (ed.), Microprocessors in Operational Hydrology, 239-245.

The computer program was written in BASIC language, in an interactive fashion. The results of calculations were plotted on graphic display and external printer.

BASIC EQUATIONS OF UNSTEADY FLOW

The flow in rivers is assumed to be one dimensional and the dependent variables are the streamflow rate, Q, and the local water level, Z, /or depth/. The flood plain is assumed to contribute only to storage.

Thus the well known continuity and the momentum equations are [3]:

$$\frac{\partial Q}{\partial x} + B \frac{\partial z}{\partial t} = q \qquad (1)$$

$$\frac{\partial z}{\partial x} + \frac{1}{2g}\frac{\partial}{\partial x}\left(\frac{Q}{F_s}\right)^2 + \frac{1}{g}\frac{\partial}{\partial t}\left(\frac{Q}{F_s}\right) + S_f = 0 \qquad (2)$$

where B_s is the width of the main channel, B is that of the entire channel /including flood plain/, F_s is the cross sectional area of main channel, S_f is the friction term and q is the lateral inflow.

FINITE DIFFERENCE SCHEME

After linearizing Eqs. (1) and (2) a four-point implicit finite difference procedure [3] was employed. The scheme is unconditionally stable and its accuracy is satisfactory for the present type of problems.

SOLUTION OF THE LINEAR ALGEBRAIC EQUATIONS

The finite difference form of the Saint-Venant equations can be written for any pair of computational points. If there are N grid points along the longitudinal axis, x, one can write 2/N-1/ equations for 2N unknowns. As two boundary conditions are also available, actually a system of 2/N-1/+2=2N algebraic equations for 2N unknowns is resulted in. This system can be solved and unknown variables at time level /t+Δt/ can be obtained.

For the solution of the system of algebraic equations the "double-sweep" method [3] was employed. The basis of this technique is the recognition that /a/ the solution is known at two boundary points of the river stretch and /b/ at any internal points the solution linearly depends on the boundary conditions.

This way an explicit type, two-step procedure can be realized for solving the system of linear equations. The major advantage of the algorithm is that the number of elementary operations, and consequently the computer time, and the RAM capacity required is proportional to the number of grid points N rather than to N^3 which is the case when using standard matrix inversion techniques.

EXTERIOR BOUNDARY CONDITIONS

They can be given in the form of Z=Z/t/, Q=Q/t/ or Q=Q/Z/ for any of the two boundaries.

INTERIOR BOUNDARY CONDITIONS

Reservoirs, inflows etc. belong to this category and can be easily handled as shown in [3].

INITIAL CONDITIONS

Z/O,x/ and Q/O,x/ are derived from a subroutine computing steady state flow.

APPLICATION OF THE MATHEMATICAL MODEL TO PRACTICAL PROBLEMS

A valuable agricultural area of about 12 km^2 in the triangle of Zagyva River Hungary, and its tributary has been saved from medium floods by the transfer of the mouth of the tributary. Additionally a flood protecting dyke along the lower stretch of tributary and another dyke /with total length of 12 km/ along the Zagyva River were constructed /Fig.1./.

Hydraulic calculations were performed to study and answer the following questions:

- what is the possible effect of dyke system /built in 1967, see Fig.1.b/ and the transfer of the mouth of tributary /see Fig. 1.c/ on flood level;
- what further rise in design flood level can be expected due to the exclusion of a part of the flood plain /which is then utilized for agricultural purposes/;
- how to determine the "optimum" crest level of the new dyke system /see Fig.1.d/ letting the extreme floods to surpass it and inundate the previously protected area /this way, an emergency reservoir could be created/.

In order to analyze the problem a data base has been established using the past and altered geometry of the affected river reaches and flood plains, the vegetation conditions and the observed flood levels. The upstream boundary condition contained the input discharge time series, while the downstream boundary condition consisted water level time series of past flood waves /the above conditions have been selected because the main problem was to determine the flood level changes at the upstream boundary/.

The mathematical model was first calibrated for various past periods by using the time series of the recording gauge "B"/Fig. 1./ situated between the upstream and downstream boundaries of the investigated river reach.The objective of calibration was to find the Manning roughness coefficients resulting in the best fitting between model simulations and measurements. These coefficients could not be directly calculated because no steady water level observations were conducted in the field. An illustration for this step of the analysis is given in Fig. 2.

Following calibration the model was validated by using independent observations as compared to the calibration phase. For design purposes a "critical" flood wave /actually monitored/ characterizing the present situation was selected. This "scenario" was employed for both, the original "natural" and the existing conditions. Comparsion of the results /Fig. 3./ gives an impression on the effect of human intervention on flood levels.

Subsequently the effect of enlarging the protected flood-plain area and higher dyke crest level was analyzed. Finally the degree of gauge-height decrease was determined for the case when the flood-plain is used as an emergency reservoir /see Fig. 3./.

RESULTS OF CALCULATIONS

Simulations of the validated model agree fairly well with the observed hydrographs /see Fig.2., Δt=4h was employed/. Model results for various situations clearly demonstrate the influence of regulation on the flood level /see Fig. 3./:

- the effect of the tributary transfer and the flood dyke on flood levels /at the gauge "B"/ is an increase of about 45-60 cm;
- the higher crest level and construction of a protected flood-plain can further rise the flood level by 15-20cm but this increase can be entirely eleminated by using the protected flood-plain as emergency resorvoir.

CONCLUSION

Based on advanced numerical methods a software was developed for microcomputers for calculating unsteady river flow. The application of this software allows to determine design flood levels for different water projects and to evaluate alternative projects for a given purpose. The software outlined here and others similar in character are expected to have a growing interest in the hydrologic practice in the future. The major reasons are the extremely fast propagation of microcomputers outside the scientific world, the friendly nature of such machines and economic efficiency.

References

1. M.Kozák: /1977/ Computation of unsteady flow /in Hungarian/ Akadémiai Kiadó, Budapest, 1977

2. J.A.Ligett J.A.Cunge: /1975/ Numerical methods of solution of the unsteady flow equations Water Res. Publ. 1975 Vol.1.

3. K.Mahmood, V.Yevyevich: /1975/ Unsteady flow in open channels Water Res. Publ. Vol.1.

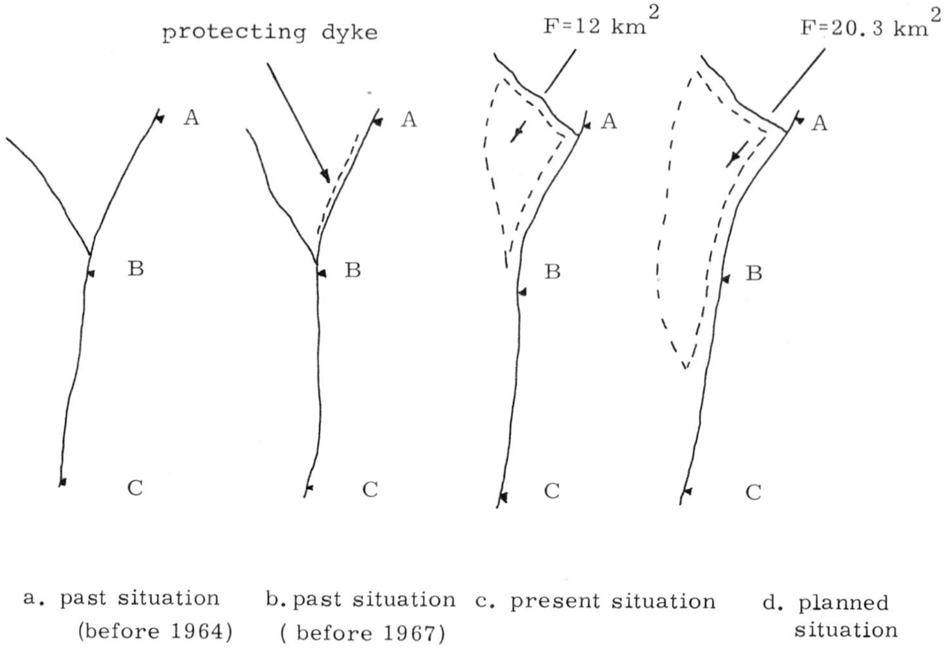

Figure 1. Studied reach of Zagyva River

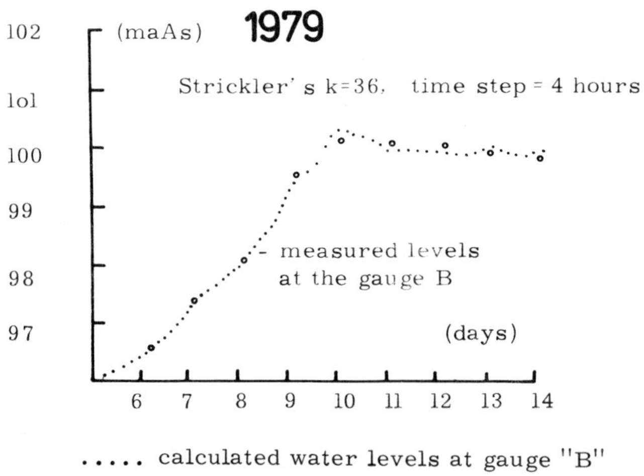

Figure 2. Calibration of the model

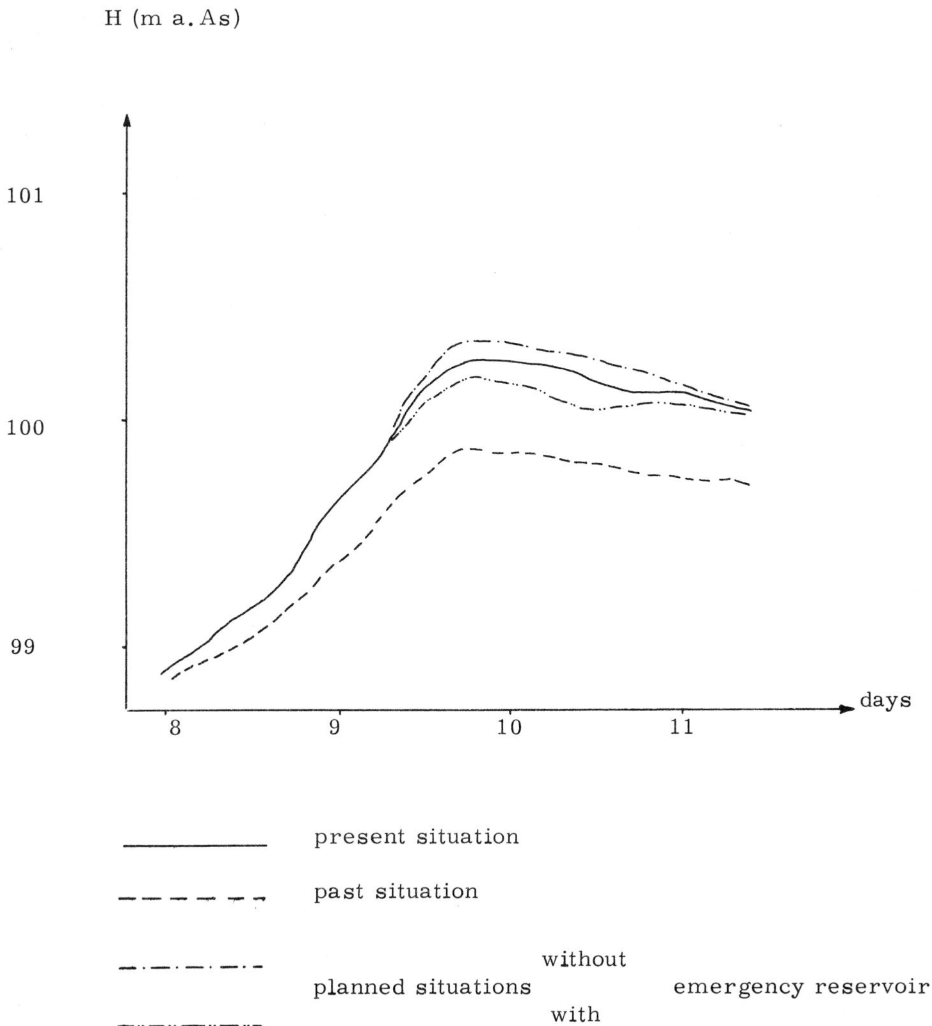

Figure 3. Calculated flood levels of different situations at gauge "B"

INDEX